the HEALING GARDEN

the HEALING GARDEN

Herbs for Health and Wellness

*A Guide to Gardening,
Gathering, Drying, and Preparing Teas,
Tinctures, and Remedies*

Deb Soule, Founder of Avena Botanicals
Photographs by Molly Haley

PRINCETON ARCHITECTURAL PRESS · NEW YORK

In gratitude to the healing
plants, to all the hummingbirds
who span the Americas,
and to the quiet ones who save
seeds and sing the sun up
each morning.

Contents

With respect, I acknowledge that Avena Botanicals herbal apothecary and farm is situated in the homeland of the Wabanaki people (Penobscot, Passamaquoddy, Mi'kmaq, Maliseet, and Abenaki). For more than fifteen thousand years, Indigenous communities have lived, and still live today, on these ancestral lands. With this acknowledgment I recognize the cruel and devastating legacies of European settlers and colonialism and commit to expanding my understanding of the intersectionality of social, racial, and environmental justice; ecological health; and land rematriation. I stand with people and organizations committed to dismantling racism, patriarchy, and all forms of oppression.

Preface

This book is more than information about herbs. The words and images are prayers on a page. It is meant to be left open in your garden basket or medicine-making room as you work. The practical information offered here comes from forty-five years of my apprenticing with herbs, of being a humble student of soil, seeds, and plants. I teach, write, and garden as a healing practice for myself and as encouragement for others to create respectful and joyful relationships with plants, trees, and pollinators.

In the garden and in wild places, the plants have been my greatest teachers. They generously offer me inspiration and instruction. Never have plants given up on me, even when people have. Standing on behalf of medicine plants, so that people may know how to grow, prepare, and consciously work with them, is worthy work. For centuries, Black, Brown, and Indigenous people and women worldwide have risked their lives (and still do) to preserve and pass on their plant knowledge. Herbalists, gardeners, and farmers need access to land and clean water, good-quality food, medicine plants, and the tools and skills to serve our communities. There is a lot of work ahead to ensure that the food and health needs of marginalized communities are met in just and equitable ways.

During the writing process, my inner landscape was sculpted by water and wind, fear and courage, confusion and grief, forgiveness and

love. Collaborating on this project with photographer Molly Haley,
a gifted artist and kind human being, nourished my spirit deeply—
thank you, dear friend. In May 2019, when Molly arrived to begin
photographing, a small hummingbird was recovering in my hand.
Throughout our work together, hummingbirds guided and delighted
us. Here our respective images and words are woven together like
a basket full of medicine plants, carefully gathered and prayed over.

This book is a gesture of gratitude to plants and pollinators, to
herbalists and healers throughout time, and to all who are my teachers.
As a rural-raised, college-educated, queer, white-identified woman,
I acknowledge that what is written here, from my experiences and
perspectives, is limited. I apologize if any of my words create hurt or
diminish another gardener or herbalist's experience.

May the healing plants infuse your heart and hands with their
spirit, guidance, and love. May they be medicine for your soul, softening your edges and inspiring you to act in ways that support safety,
peace, and harmony for all living beings.

Deb Soule, Spring 2020

Daylilies and valerian
flowers growing
beneath one of Avena's
silver maple trees, with
Avena's greenhouse in
the background

Part I

GATHERING WITH GRATITUDE

LEFT
A sign near the entrance
to Avena Botanicals
apothecary

ABOVE
Greeting and giving
gratitude to the
hawthorn trees as I
enter the garden

Respect. Reciprocity. Gratitude. Humility. Love.

I was a child when my grandmother whispered to me these words, directly transmitted without any explanation. After many years of living and learning, attempting to be a good human being and making repeated mistakes, I find that my grandmother Katherine's teachings have begun to infuse my heart and hands. They guide my work as an herbalist, gardener, and teacher. I feel most grounded and humbled when my hands and fingernails are consistently stained brown from soil, when I have had enough quiet time in the garden or woods to center my soul and spirit.

I was taught as a teenager to respectfully approach and address a plant before ever touching or picking the plant. If permission to gather is granted, then leaving a gift, an act of reciprocity and respect, is what follows. At the entrance to Avena Botanicals' apothecary is a hand-painted sign that says "Reciprocity of Pollination." It is placed intentionally to inspire visitors, students, and staff to pause and contemplate what these words point to: a world that values nature and recognizes how essential pollinators are to life. A balanced world is one rooted in generosity and reciprocity.

Connecting and Communicating with Plants

Plants are living beings. They are intelligent and communicate in caring and complex ways with all who live in their ecosystem. Cultivating a sense of humility and a feeling of warmth and respect for plants is vital for creating meaningful relationships with them. My grandmother showed me that plants appreciate the effort humans make through the slow, quiet, and regular way she visited various apple trees and spring flowers year after year.

If you are new to communicating with plants, begin by sitting or standing near a plant or tree as a regular practice before touching, tasting, or gathering. Introduce yourself to the plant or tree. As awkward as this may seem at first, over time and with sincere effort, communication and connection with plants can develop and deepen.

Give yourself time and space to be with plants. This is in itself a healing practice. Let the plants inspire you. Keep a journal with you. Place your hands over your heart and breathe, letting your breath help you stay focused, grounded, and curious. When your mind wanders or if doubts, criticisms, or fears slip in, let your breath bring you back again and again to the present moment. You are developing a heart-centered approach to listening and communicating with plants. When your communication feels complete, depart consciously and respectfully. I often bow and speak aloud my gratitude and farewell.

When approaching plants you wish to collect for medicine, sit, kneel, or stand near them. Quiet yourself by taking some deep breaths and allow yourself time to hear and feel life pulsing under and around you. Slow down. Once you feel grounded, begin a conversation with the plants. Open your heart and sincerely ask if they are willing to share their gifts with you. Let the plants know why you have come, why their medicine is needed. Even when collecting an annual herb such as calendula several times throughout the summer, attune yourself to the spirit of the plant each time you approach. When permission is given, offer a gift, something that has genuine meaning for you—a song, a prayer, a poem, a piece of your hair, a small bit of food you have made. If permission is not given, trust the mystery of this moment.

Remember, you are not alone as you gather herbs. Birds, butterflies, insects, other animals, and elemental beings may be nearby, sometimes

Before every harvest, each gardener places a little bit of pearly everlasting at the base of the plant as an offering of gratitude.

visible and sometimes not. Nature spirits may also be present. Many Indigenous peoples around the world acknowledge the presence of elemental beings and nature spirits at work in their gardens, fields, forests, shrines, temples, sacred wells, and healing centers. We see this in places such as Ecuador, Nepal, Tibet, India, Bhutan, Sri Lanka, Bali, and Japan. In Scotland, Ireland, and England, certain people perform rituals that honor elemental beings and nature spirits, the natural rhythms of the sun and the moon, and the sacredness of water and stones. In North America, some Indigenous people recognize the Little People, and most honor seasonal, solar, and lunar rhythms.

Collecting herbs offers the opportunity to practice mindfulness. Bring yourself into the present moment. If you are with other people and conversations arise, keep the energy and the content positive and in service to the plants you are gathering. Singing, humming, and laughing when alone or with other people can be wonderful ways to exchange energy with plants.

BELOW
Spending a moment with a meadowsweet plant

OPPOSITE
A native Maine bee pollinating one of Avena's lavender plants. Avena is a sanctuary for pollinators.

Connecting with Place

My life as an herbalist and gardener has long been rooted in the rhythms of the natural world. Growing up in rural Maine awakened my curiosity and respect for Earth's life-affirming forces and seasonal shifts, and the ever-changing cycles of the moon. For the last thirty-five years, I have had the privilege of living on land in the same community, allowing me to observe where on the horizon the sun and the full moon rise and set throughout the year.

I also acknowledge that I live in Wabanaki territory, home to the Penobscot, Passamaquoddy, Mi'kmaq, Maliseet, and Abenaki peoples, on land that was never ceded to European settlers. At the beginning of European contact, between 1616 and 1619, three-quarters of the Wabanaki people died. I speak this truth because this place I feel connected to has a brutal and unhealed history that is barely being addressed. (See Social, Racial, and Environmental Justice Resources, page 214.)

Following Seasonal, Solar, and Lunar Rhythms

Depending on where you live, seasonal transitions may be pronounced or subtle. In Maine, we have five distinct seasons (winter, spring, summer, late summer, and fall). Closer to the equator, seasonal shifts are less obvious. Regularly noticing and aligning oneself with nature's rhythms, wherever you live, helps to restore a feeling of connectedness and vitality and enhances your ability to be sensitive and discerning; to make personal and collective choices that foster inner peace and compassion; and to cultivate consideration, equality, and safety for our Earth and for all people.

Avena's garden soil is healthy and alive with organic matter and millions of microorganisms.

Soil

Soil is the precious skin that covers the body of Earth, offering nourishment and protection, and supporting plants and trees to take root, grow, and thrive. Skillful gardeners and farmers are devoted to soil. If you seek a holistic, ecological, and spiritual agricultural model that moves beyond the industrial organic model, consider studying and implementing biodynamic practices. Biodynamics recognizes that soil is alive, breathes, pulses, and has an intelligence and spirit. Take time to regularly observe and nourish your garden soil. Medicinal plants, and all life, depend on healthy, nontoxic soil (earth), water, air, and warmth to flourish physically and spiritually.

Biodynamic Gardening

Biodynamic agriculture informs my understanding of seasonal, solar, and lunar rhythms; the twelve zodiacal constellations; and their relationship to plants. Biodynamic calendars designate the most favorable

October 2019

times for planting root crops, leaf crops, flowers, and fruits and are based on the moon's daily astronomical movement through these twelve constellations.

Sherry Wildfeuer has compiled and published the *Stella Natura Biodynamic Planting Calendar* in the United States annually since 1978. She was inspired by the work of Maria Thun, a biodynamic practitioner from Germany who in the 1950s began recording her observations of how the moon's monthly movements affect seed germination and plant growth. Wildfeuer writes:

This calendar is meant to be used with common sense and an eye to the weather.... The astronomical information offered in the [monthly] charts is, like the weather, part of the nexus of environmental factors which affect the plants in your care. If you wish to support their growth by creating optimal conditions, you will want to pay attention to this information— and to the weather.

On Avena's farm, I have been using the *Stella Natura* calendar since 1986 to guide our seed planting and transplanting, biodynamic

A page from the *Stella Natura Biodynamic Planting Calendar*, showing my work notes scheduling Avena's harvests for black cohosh, teasel, and echinacea roots

Biodynamic Agriculture

The principles of biodynamic agriculture were first formally outlined in a series of eight agricultural lectures given by Dr. Rudolf Steiner in Koberwitz, Silesia, June 7–16, 1924. Steiner was speaking to a group of farmers who had repeatedly approached him for advice regarding the decreased vitality of their seeds, soil, and animals. The farmers' observations coincided with the end of World War I, when leftover munitions were beginning to be used as agricultural chemicals. These eight lectures, given to farmers familiar with Steiner's spiritual cosmology (known as anthroposophy), became the foundation for the preparations and practices now known as biodynamics.

Steiner believed that the environmental damage caused by chemical fertilizers, toxic pesticides, and intensive farming methods would not be easily remedied. His ideas for supporting the Earth's health first formed in his youth in rural Austria, where he grew up around people who followed the rhythms of nature and relied upon herbal medicine for healing. Steiner's subsequent formal education included studying the philosophy and scientific work of the German writer Johann Wolfgang von Goethe. He was also influenced by the herbalist and healer Felix Koguzki, who spent his life serving people and animals, often carrying bundles of herbs on his back for many miles to reach the train into Vienna.

Biodynamic gardening and farming relates the Earth's ecology to the larger cosmos. It puts significant focus on soil health, viewing the soil and farm as living organisms, and works with seasonal, solar, and lunar rhythms. Though sometimes challenging to express in words, biodynamics recognizes and nurtures the spiritual presence and unseen life forces in gardens, farms, fields, and forests. Biodynamics is practiced worldwide and continues to be shaped by the ecological, spiritual, and cultural practices of people and place. (For more information on biodynamics refer to Resources, page 214.)

preparation applications, and root-digging days. This and more of my favorite biodynamic and gardening books are listed in the Resources (page 214). Michael Phillips's *Mycorrhizal Planet* is an excellent resource for understanding soil biology, and it has a great bibliography. My book *How to Move Like a Gardener* includes an easily understandable chapter on biodynamic gardening.

The Four Elements, the Constellations, and Plant Parts

For centuries, each of the twelve zodiacal constellations has been associated in the Western world with one of the four primal elements: earth, water, air, or fire. The information below connects these twelve constellations and four elements with particular parts of the plant (root, leaf, flower, and fruit/seed).

Constellations aligned with the earth element

TAURUS, VIRGO, CAPRICORN

Roots are associated with the earth element. The seeds of root crops such as black cohosh and teasel are planted when the moon is moving through Taurus, Virgo, or Capricorn. We also dig roots for medicine making when the moon is moving through these constellations.

Energetic qualities of roots: grounding, rooting, reconnecting

Constellations aligned with the water element

CANCER, SCORPIO, PISCES

Leaves are associated with the water element. The seeds of leaf crops such as lemon balm and holy basil are planted when the moon is moving through Cancer, Scorpio, and Pisces.

Energetic qualities of leaves: movement, flexibility, fluidity, freedom

A frog in Avena's pond

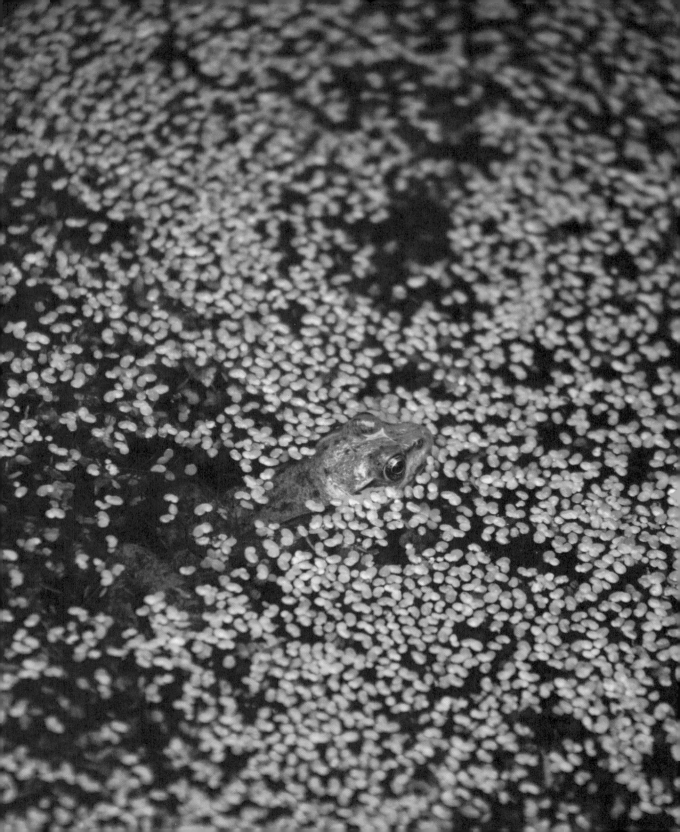

Constellations aligned with the air element

GEMINI, LIBRA, AQUARIUS

Flowers are associated with the air element. The seeds of flowers such as calendula and wild bergamot are planted when the moon is moving through Gemini, Libra, and Aquarius.

Energetic qualities of flowers: inspiration, expansiveness, transformation, soul shifts, beauty, light

Constellations aligned with the fire element

ARIES, LEO, SAGITTARIUS

Fruits and seeds are associated with the fire element. Oat, tomato, and zucchini seeds are planted when the moon is moving through Aries, Leo, and Sagittarius.

Energetic qualities of fruits and seeds: carriers of memory, manifestation, fruition, initiation

Digging Roots

All herbs have a specific moment in their life cycle when their medicine is strongest. Many herbalists dig roots in autumn, once the plants' seeds have matured and their tops have died back. Autumn is when the plants' energy and nutrients return to the roots and become concentrated there. Roots can also be dug in spring, when their first leafy shoots poke above the soil line. Once a plant begins producing leaves, however, the nutrients move up into the stalks to support the development of leaves and flowers, and the roots' medicinal properties become less concentrated. On Avena's farm, we mostly dig roots in autumn, after the plants' ripened seeds have been collected and saved for the following year's plantings.

Harvesting black cohosh root in the fall

(See part 4, "Healing with Herbs," for specific gathering guidelines for individual herbs.)

I follow the biodynamic planting calendar when digging roots. The ideal lunar phase for digging roots and planting root crops from seed or root cuttings is when the moon is moving through the constellations associated with the earth element: Taurus, Virgo, and Capricorn. From a gardener's perspective, this makes sense—roots grow in the earth and are aligned with the earth element. Digging and washing roots invites us to slow down and explore the hidden realms of plants and soil. From an herbalist's perspective, drinking a root tea or taking a medicinal root tincture can be beneficial when the energetic qualities of roots—grounding, rooting, reconnecting— are needed for supporting emotional and spiritual balance.

CLOCKWISE FROM TOP LEFT
Digging black cohosh root with a fork during fall harvest

Harvesting teasel root with a sturdy garden fork

Gardener Anna Leavitt collecting black cohosh and blue cohosh seeds in the fall

Tools for Digging

On Avena's biodynamic farm, we dig our roots with either a sturdy garden fork or a strong, narrow spade. Early in my root-digging days, I broke a fork while digging roots. This experience guided me to invest in high-quality stainless-steel tools with well-made wooden handles. High-quality tools last for years if kept oiled and wiped down and stored indoors. Early in spring or at the end of the gardening season, we oil the wooden handles with a blend of half turpentine and half raw linseed oil, applied with a clean cloth. (See page 216, Seeds and Tools.)

The Digging Process

Keep your back and body strong and flexible and better prepared for digging roots through regular stretching, yoga, dance, or qigong. Smaller roots come out of the ground easily, while larger and older roots, such as marshmallow and comfrey, require more coaxing and patience. Do your best to loosen the soil around each plant so you can remove the whole root without breaking it; a bit of leverage with a fork or shovel, or two people working opposite each other, may help.

Caring for Soil and Roots

It is best to dig roots when the soil is not too wet and muddy. Dry soil is less compacted, and its structure less damaged, so digging is easier. And dryer soil is more easily shaken off the roots, leaving more of the precious soil in the garden. Take the extra time to work the soil out of the roots with your fingers. Larger root clumps, such as echinacea and valerian, can be broken apart in the garden with a sharp spade or pulled apart by hand, so the soil embedded in the clumps can be shaken free and returned to the garden bed. Melanie Carpenter and

Jeff Carpenter of Zack Woods Herb Farm in Vermont, large-scale organic herb growers, use a sharp field knife and rubber mallet to break apart dense roots before washing.

Once you've dug a bed of roots, take a shovel or rake and refill all the holes with soil. This is a good time to weed the bed thoroughly. Add a few inches of compost on top of the soil, mix the compost and soil gently while raking, and then mulch the bed with two to four inches of straw (or seaweed, if you live near the ocean). Covering bare soil with mulch prevents erosion, and over time, the mulch naturally decomposes and adds valuable organic matter.

During the harvesting process, we place dug roots into food-grade plastic buckets. Designate a medicinal root bucket and/or wheelbarrow that is never used to haul manure and compost. Once we fill a bucket with harvested roots, we tuck it into a shady spot or carry it to our root-washing station while the digging continues.

Washing Roots

The less soil embedded in the roots, the easier and quicker it is to wash them—another reason to remove excess soil before the roots leave the garden. At Avena, the outdoor water spigot we use for washing roots is plumbed with hot and cold water (and is attached to a lead-free hose), allowing for a lukewarm temperature that is pleasant on our hands, especially on chilly autumn days.

We first rinse the roots a few times in five-gallon buckets by filling the buckets with warm water and agitating the roots by hand. This wash water can be poured onto the bases of plants or trees that need water or placed in twenty-five- or fifty-gallon food-grade plastic containers for future waterings.

CLOCKWISE FROM TOP LEFT Harvesting teasel root

Freshly harvested and washed teasel root wrapped in a cloth

Gardener Brittany Cooper washing teasel root in a wheelbarrow filled with a mix of hot and cold water

Food-grade plastic bucket designated only for root harvest

We then transfer the roots into a yellow wheelbarrow and use a spray nozzle to remove the more firmly attached soil. We pull the roots apart by hand and cut them with Felco heavy-duty hand pruners (#6, #8, or #9). The number of times we wash them in a bucket and/or wheelbarrow depends on the structure of the roots and how much soil still remains. Another method is to lay the roots on wire screens, hose them down using medium pressure (overly high pressure will damage them), and let them drip-dry for an hour before making them into a fresh tincture or chopping and drying them for a future tea or salve.

Chopping and Drying

The best time to chop roots is soon after washing. Let the water drip from them for an hour or so on a wire screen outdoors or in a green-house, then chop them and lay the pieces on drying racks. We make pieces that are ½–1 in. (1.3–2.5 cm) in size. Any rotten or woody parts are discarded into a compost bucket. A clean cutting board and sturdy knife or Felco heavy-duty pruners work well for chopping small or midsize roots.

Collecting Leaves and Flowers

The aromatic oils and medicinal properties of leaves and flowers are strongest when collected on sunny mornings before the sun's rays wilt them. Look for leaves and flowers that feel vibrant and full of life. Avoid ones that are bruised, discolored, or full of insect holes.

Every herbalist has a preference regarding when and how to gather and prepare herbal medicines. I like to gather roses and lady's mantle leaves and flowers for making tinctures in the early morning, when birdsong and morning prayers infuse the garden with joy and lightness.

The early-morning dew still present on the plants adds a touch of magic to the tincture-making process. When gathering roses and lady's mantle flowers and leaves for drying, on the other hand, I wait until midmorning before collecting and laying on screens to dry because the dew hinders the drying process; best to wait until the dew has evaporated.

Carefully placing leaves and flowers into a beautifully crafted basket honors their medicine and spirit. Woven baskets allow airflow and prevent herbs from wilting or deteriorating too quickly. Plants harvested into plastic bags or piled up too high in any kind of container will sweat, bruise, and lose their medicinal properties.

LEFT
Tools for the harvest: Felco 310 clippers, a gratitude-offering pouch, and an assortment of herb-harvesting cloths

RIGHT
Freshly harvested lavender

Pay attention to the intensity of the sun when picking herbs. You may need to keep a cotton cloth on top of your basket for shade. When gathering several baskets of herbs, I keep all but the one I am using in the shade of a nearby tree. Collecting into a clean brown paper bag devoted exclusively to herb gathering also works. Keep your paper bags in a special place and label them "Herb-Gathering Bag."

Leaves

The water activity in a leaf or stem contributes to the various chemical changes that occur in plants, photosynthesis being one of these amazing, complex processes. Leaves produce food through photosynthesis, and this food is transported throughout the stem and leaf structure. Closely observing a perennial plant or tree's leaves unfurling and developing year after year enhances an herbalist's understanding of when specific medicinal leaves are most potent for picking.

Felco 310 clippers work well for harvesting most leaves. I use an Asian tool called a *kama* when gathering large amounts (more than 5 lb. [2.3 kg]) of nettle and lemon balm stalks and comfrey leaves. I place them onto a large cloth, and once it's filled, I carry the cloth to the drying room and carefully lay the leaves on drying screens.

You will learn the fine details of when each individual herb's medicine is strongest from experience, books, research, and exchanges with other herbalists. For example, the leaves of some mint family plants such as peppermint, spearmint, horehound, and holy basil are known to contain higher amounts of aromatic oils when the plant is flowering, so wait to harvest these herbs until flowering begins. Nettle and comfrey leaf (not in the mint family) are best harvested before the plant flowers. (See part 4, "Healing with Herbs," for harvesting

ABOVE
Harvesting nettle with
a *kama*

RIGHT
Harvesting lavender
with Felco 310 clippers

information on individual herbs.) Greg Whitten's *Herbal Harvest*, Jeff Carpenter and Melanie Carpenter's *The Organic Medicinal Herb Farmer*, and my own book *How to Move Like a Gardener* are excellent resources on specific timing for herb harvests.

Flowers

While visiting Avena's farm a number of years ago, the European herbalist and author of many herb books Juliette de Bairacli Levy instructed me to gather flowers when they first open, before the bees frequent them too many times. This ensures that the flowers are of the highest quality. After many years of gathering flowers, I say the same to students. Flowers' color, texture, and fragrance are more vibrant and alive when they first open.

I gather some flowers, such as calendula, mullein, red clover, and nasturtiums, with my fingers. For hawthorn, lavender, and roses, I use Felco 310 clippers. Be mindful to pile flowers carefully when gathering, and shade the basket on hot, sunny days. Lay the flowers on screens to dry soon after your harvest is complete. If fresh flowers are being prepared to make a tincture, vinegar, tea, or oil, process them immediately after you come indoors.

Picking Medicinal Fruits and Seeds

The ripening of fruits and seeds depends on weather and soil health, requiring regular—sometimes daily—checks on changes in their color, taste, and texture to know when they are ready to pick. Pulpy fruits such as *Schisandra* and elderberries are harvested when ripe. Their taste and color lets me know when they are ready; if I wait too long, they ferment on the vine. *Schisandra* berries and elderberries grow in clusters. I harvest them using Felco 310 clippers and lay them in baskets lined with a cotton cloth. If the summer is hot, with little rain, our *Schisandra* berries start to ripen in late August. With plenty of rain, they start ripening in early September.

With oats, the green, milky seed stage is the time for gathering for medicine. The timing of this harvest depends on when in late spring or early summer the seeds were sown and how warm, wet, or dry the weather has been during their growing cycle. I begin checking for milk, squeezing several seeds daily, when the green oat seeds first start to noticeably swell. The patch is ready for collecting when milk emerges from not just a few but numerous seeds. There is a window of about four to seven days when the milk is perfect and we gather the seeds for tincturing or drying. Oats are easily harvested by hand-stripping the seeds off their stalks and placing them in a tightly woven basket or brown paper bag.

Small seeds such as nettle demand more patience and attention than larger seeds and fruits. When the tiny, greenish nettle seeds start to turn brown, the whole stalk can be shaken into a brown paper bag to release the seeds. Small seed heads such as chamomile, valerian, and holy basil can be clipped into a bowl or paper cup. Larger seed heads such as echinacea or angelica can be cut into brown paper bags.

Meditation

Meditation is an invitation to rest the mind. Whether you are engaged in a formal Buddhist practice or in other forms of meditation or prayer, regular meditation slows down reactive patterns of the ego and offers guidance for living more fully in the present. Science shows that meditation slows brain waves; as this occurs, we sense spaciousness and

LEFT
Milky oat seed (Latin name *Avena sativa*, from which Avena Botanicals gets its name)

ABOVE
Sitting on a bench near a tiny pond, I watch ruby-throated hummingbirds frequent the red bee balm and white nicotiana flowers.

relaxation. The body and mind can heal when we feel more relaxed. People who have experienced trauma and are interested in meditation may refer to David A. Treleaven's *Trauma-Sensitive Mindfulness* and Laurence Heller and Aline LaPierre's *Healing Developmental Trauma*.

Sitting quietly, even for five minutes daily, eases the busyness of doing that agitates the mind, especially for gardeners and farmers during the growing season. Put a bench or chair in your garden for meditating during warmer months. Drinking tea, gently touching and smelling herbs, walking slowly in the garden, watering seedlings, stirring biodynamic preparations—these are all opportunities to practice mindfulness. Coming back to the breath, again and again, grounds us in the body and in the present moment. Even pausing to take just three long breaths helps reset an unsettled nervous system and a mind run amok.

Many years ago, I began repeating individual words while planting seeds as a way to stay present, to generate a feeling of love and warmth in my heart, and to infuse a generative feeling like compassion into the seeds and spirit of the plants to come. Words I commonly use include *compassion*, *kindness*, *joy*, *peace*, *gratitude*, and *love*. Seed by seed, love can be cultivated in our hearts.

Wildcrafting Herbs

Herbalists and healers have practiced foraging—gathering wild edible foods and herbs from places or spaces that are not actively being cultivated—for centuries. While this book's focus is on growing and gathering herbs from gardens, I briefly mention wildcrafting because of its increasing popularity. A few key challenges to consider include:

❦ overharvesting of wild herbs
❦ lack of education regarding how to responsibly and respectfully gather plants
❦ lack of knowledge of at-risk and endangered plants (see United Plant Savers [unitedplantsavers.org] and state organizations for lists of endangered plants)
❦ lack of understanding of the ecological significance and challenges native and non-native plants and Indigenous people face in specific ecosystems due to environmental toxins, loss of habitat, and climate change
❦ entitled attitudes, namely the belief that anyone should be able to wildcraft herbs freely without first investigating who may already be collecting plants from the area

I was taught as a young person never to collect the largest plants, but to honor these as the grandmothers. Always ask permission, leave offerings, collect only what you need, gather from places where the plants are abundant, take less than 25 percent of the plants, and return year after year to observe the health of the plants and the place. Are the plants proliferating or depleted? Does the area need weeding or pruning? Do aggressive non-native species need removal so the native plants can thrive? Has the area been sprayed with toxic chemicals, or has it ever been used as a dump site?

    Our lives as humans are intertwined with the land, water, plants, and animals as well as the people who came before us. Take time to learn the history of the Indigenous people of the land where you live, garden, or wildcraft. What happened to their original lands? How many Indigenous people are alive today, and where are they living?

How many speak their language fluently? Do you live on or near land that was once worked by slaves? How will you commit to better understanding decolonization and the intersectionality of racism, sexism, classism, and other forms of oppression that contribute to suffering and injustice? (See Social, Racial, and Environmental Justice Resources, page 214.)

Planting and tending medicinal and food gardens for pollinators and people is an act of resistance against the oppressive patriarchal systems that perpetuate the medical- and agricultural-industrial complexes. Communities everywhere need access to clean, healthy, safe garden spaces where people can grow food and herbs. If you have a garden, invite your neighbors in. Share the harvest. Get involved with others to create community medicine gardens in your neighborhood or city block. Build relationships as you build healthy soil. Share skills, seeds, seedlings, space, and songs.

> The moral covenant of reciprocity calls us to honor our responsibilities for all we have been given, for all that we have taken. It's our turn now long overdue. Let us hold a giveaway for Mother Earth, spread our blankets out for her and pile them high with gifts of our own making.... Whatever our gift, we are called to give it and to dance for the renewal of the world. In return for the privilege of breath.
> —Robin Wall Kimmerer, *Braiding Sweetgrass*

LEFT
A monarch caterpillar feeding on milkweed

BELOW LEFT
A monarch chrysalis attached to a mountain mint plant

BELOW RIGHT
A monarch chrysalis attached to a recently harvested lemon verbena leaf. This chrysalis was moved to a place where it could safely continue to transform into a butterfly.

ABOVE LEFT
A monarch chrysalis that has turned black and translucent means that the butterfly is almost ready to emerge.

LEFT
When the butterfly emerges, its wings are wet and curled and it must wait several minutes or hours for its body to dry before it can fly.

ABOVE RIGHT
A monarch butterfly pollinating a Mexican sunflower

Part II

DRYING HERBS

How you dry herbs is as important as when and how you gather them. The main purpose of drying herbs is to remove moisture and, very importantly, to preserve their medicinal (bioactive) properties and vitality (spirit) for your teas, tinctures, oils, salves, steams, baths, and food. Herbs dried skillfully and mindfully retain the vibrant color, fragrance, and flavor of when they were gathered. With experience, you will learn how each specific herb feels, smells, and looks when dried carefully, so keep good notes. Let this be a practice of gratitude, especially when you feel overwhelmed by life's daily tasks. The following guidelines for obtaining high-quality herbs come from my years of drying herbs, both in my home and in Avena Botanicals' drying rooms.

The Drying Room

Be creative with the spaces you have available for drying herbs. Choose one that is warm but not too hot, such as an attic in northern climates, as long as its temperature is not above 100°F (38°C). In my first home, drying racks attached to the ceiling, where warm air hovered, functioned fairly well. I prevented dust from settling on the herbs by laying thin cotton muslin cloths over them. Fans for circulating air support the drying process. Window shades are important for protecting the plants from sunlight, which fades their color and lessens their potency. Avoid spaces where rodents roam. (See page 59 for how we created rodent-proof drying rooms in Avena Botanicals' 1830s barn.)

Learning to manage your drying space and the nuances and needs of each individual herb is a process: pay attention and write down your observations. Have fun, laugh a lot, and share what you learn with others.

Temperature (Hot-Air Drying Method)

Bring your herbs into the drying room before noon, especially on hot summer days. The ideal temperature for drying herbs is 75-100°F (24-38°C). Preferably, start the drying process at a temperature at the lower end of this range. Heat above 105°F (41°C) begins to destroy the essential oils and other valuable constituents. High temperatures not only diminish the medicinal qualities but also dry the outermost cells first, leaving the herb unevenly dried. I experienced this firsthand one summer, losing dozens of pounds of milky oat seeds because the inner moisture had not fully evaporated. Thinking the oat seeds were fully dry, I bagged them up and a month later found they had molded.

Avena has a small Rinnai propane gas heater in one of our drying rooms that we only use on cool, rainy days in spring or fall. Our two drying rooms are located on the second floor of our barn, where the natural heat of the Maine summer is all we need. One room is insulated, and the other is not. The former has a regular door and a screen door made with metal hardware cloth. If the temperature rises above 100°F (38°C) in the insulated room, we open the door to let the excess heat escape while the screen door keeps rodents from sneaking in.

A thermometer, humidity gauge, and notepad for recording the drying room's temperature and humidity level are all helpful. It is interesting as well to make daily notes of the outside temperature and weather to learn how weather patterns affect the drying room's humidity.

Air Circulation

Mounted or standing floor fans ensure good air circulation. As the air moves, it picks up moisture from the surfaces of the fresh herbs. Even, steady airflow throughout the drying area dries the herbs on each screen more uniformly.

CLOCKWISE FROM TOP LEFT
Checking yarrow one day into its drying process. Drying nearby are Tulsi and calendula flowers.

Checking herbs on the drying room racks

A fan and a dehumidifier are essential tools in the drying room. Here, milky oat seed is laid out to dry on multiple drying racks.

Consider using an energy-efficient dehumidifier if you are drying in a smaller, enclosed room. We keep a dehumidifier running whenever herbs are drying and empty its water-collecting tray every morning and evening into a bucket just outside. (This water can be used for watering potted plants or garden plants, trees, and shrubs.) A dehumidifier is important for smaller-scale herbalists and gardeners because we often bring fresh plants into the same drying room multiple times throughout the week, introducing more moisture, which can negatively affect the herbs already drying in the room. Research dehumidifiers to find the size suited for your drying space.

If you are drying in a shed or hoop house, ventilation fans are another method of pulling moisture out of the air and lowering the temperature. Refer to Jeff Carpenter and Melanie Carpenter's *The Organic Medicinal Herb Farmer* for details on this system.

Drying Screens and Racks

Drying Screens

Screens made from fiberglass or nylon mesh (available in most hardware stores) work well for drying herbs. Wash the mesh before use by unrolling it outdoors on a clean, grassy area, spraying it with a hose, and letting it dry in the sun. Then cut and staple the mesh onto wooden frames. If you live in an urban area, you might wash and drip-dry your screens indoors, in a shower or sink. Each spring, in preparation for the growing season, vacuum your screens or sweep them with a designated brush that you use only for cleaning herb screens. Label your brush and hang it near your screens.

The size of your frames and racks will depend on the size of your space. Be sure when designing your drying rack that you leave plenty

Freshly harvested lemon verbena laid out on a drying screen

of room for pulling the frames in and out of the racks. A nearby table is helpful for holding each screen/frame while you lay out fresh herbs and remove dried ones.

Larger Frames

We use larger frames for drying nettle, lemon balm, lady's mantle, and comfrey leaves. Our wooden rack that holds seven frames is 72 in. (183 cm) tall. The frames measure 45 × 54 in. (114 × 137 cm) each, and they sit on wooden runners made from strips of wood that are 1 × 1 × 54 in. (2.5 × 2.5 × 137 cm) each.

Smaller Frames

Our wooden rack that holds eleven smaller frames is 50 in. (127 cm) tall. The frames measure 27 × 40 in. (69 × 102 cm), and they sit on wooden runners made from strips of wood that are 1 × 1 × 40 in. (2.5 × 2.5 × 102 cm) each.

In my small, off-grid home, I use a collapsible metal clothes-drying rack to support my three smaller drying screens. This rack is upstairs in a loft area, where in summer the heat dries flowers and leaves well. Open, screened windows provide air circulation.

Collapsible Drying Racks

Fedco Seeds sells a unit of six round collapsible racks, made from a durable polyester netting with a metal frame and center strap for weight support, with an overall diameter of 36 in. (91 cm). The racks easily snap together to create a vertical herb-drying system for people living in small spaces. You can sweep or vacuum the racks. When not in use, the whole system folds up neatly and fits into a storage bag.

LEFT
Laying out freshly harvested hawthorn leaves and flowers to dry

RIGHT
Placing a screen of *Rosa rugosa* petals into a wooden rack that holds several screens

Laying Out Herbs

Lay plants loosely on screens or in a basket so air circulates among them. Singing while laying out herbs to dry is a life-affirming activity that benefits plants as well as people. We lay flowers like calendula individually on their side on the screen so that as many of their delicate petals will remain attached as possible (this is preferable, in our experience, to laying the flowers directly downward or upward facing). Each calendula flower's center is a unique mandala, embodying the warmth and light of the sun. With leaves like comfrey and basil, which blacken easily if bruised, we take extra care to lay them out individually. Herbs like holy basil, lemon balm, and nettle are laid out a bit more thickly, as the leafy stalks naturally create air spaces. Ideally each layer is no more than 2–4 in. (5–10 cm) thick. As the moisture evaporates after a day or so of drying, these layers will shrink. The woody and fibrous tissue of roots requires a longer time to dry than leaves and flowers do.

Drying in Baskets

Fresh herbs like lemon balm, hawthorn flowers and leaves, nettle leaves, and roses dry well when loosely laid into baskets made from natural materials. The weave of the basket allows for airflow. A thin cotton cloth over the basket prevents dust from settling on the herbs. Place the basket in a warm location out of the direct sun, and check it daily.

Hanging Herbs

Herbs with sturdy stems such as mugwort, yarrow, lavender, sage, and other mints can be tied into bunches using rubber bands and hung from horizontal strings, wires, or wooden beams. Be mindful not to tie them too tightly or in too big a bunch, as this will prevent the center of the bunch from drying. Once the herbs have dried, strip the leaves and flowers from the stems and store them in a glass jar. Compost or make footbath teas from the stems. Hanging bunches of herbs adds beauty to a living space, but they will collect dust and lose their medicinal properties if left hanging for more than a few weeks.

Drying Times

Drying times vary for different flowers, leaves, and roots. If you are using a rack stacked vertically with several drying screens, lay the most recent harvest on the top screens so the newly introduced moisture won't be reabsorbed by the partially dried herbs. We aim to keep the humidity in Avena's drying rooms below 62 percent during the first

RIGHT
Yarrow hanging to dry

BELOW
Gardeners laying out
freshly harvested
calendula flowers
on a drying screen

several hours that fresh herbs are drying. The dehumidifier helps pull excess moisture from the air.

Check your drying herbs daily, because some will dry faster than others. Touch and gently crumble leaves in your hand to sense if they are dry; also look at them and smell them. If a leaf such as comfrey has moisture remaining, the base will be pliable. As soon as you are confident that certain herbs are dry, remove them from the drying space. Overly dried leaves will crumble in your hand and become almost dustlike.

Of the flowers we dry, calendula takes the longest. Our records over the years show that approximately 8 lb. (3.6 kg) of fresh calendula flowers reduces to 1 lb. (0.5 kg) when dried. Check the centers of several calendula flowers to ensure they are fully dried, as the petals dry much more quickly than the centers. You can do this by cutting the center of the flower and rubbing it on your hand; you will feel any moisture still remaining. Delicate flowers such as chamomile and red clover dry more quickly than calendula and must be removed from the rack as soon as they are dry or their quality will diminish.

Roots are the only part of a plant that we wash and chop before laying on screens to dry. Roots take longer to dry than flowers and leaves, but if they take *too* long, they may develop mold or high yeast levels. Exposing more of the root's surface to air speeds the drying process. Small-scale gardeners can cut roots into ½ in. (1.3 cm) pieces using Felco #6, #8, or #9 heavy-duty pruners or a super-sturdy knife. Ideally, the room should be at least 80°F (27°C), which is hard to attain in autumn without extra heat. Drying roots above a woodstove can work, as does a food dehydrator (see Seeds and Tools Resources, page 216) set at 100°F (38°C).

A basket of freshly harvested anise hyssop

Jars labeled: Milky Oat Seed, Calendula, Hawthorn leaf + flower, Chamomile, Dandelion Root

Glass jars containing dried herbs for winter teas

Weather

Weather plays a big role in how well herbs grow, when and how they are gathered, and how well they dry. Keep weather records alongside your gathering and drying notes. Several damp, rainy days in a row will affect gathering and drying schedules.

Weather (along with pesticide and herbicide poisoning) affects the health of all plants, pollinators, birds, water, and soil. The Earth's climate is changing quickly, and pollinators and birds are seriously declining in numbers. Please create safe, chemical-free sanctuary gardens for plants, pollinators, and people in your front or back yard and in community spaces. All of us rely on plants for food, and people around the world rely on plants for medicine—many out of necessity and a growing number by choice.

Storing Dried Herbs

Airtight glass jars work well for storing dried herbs. If your shelf space for jars is limited, keep your herbs in a clean brown paper bag placed inside a clean, thick plastic bag closed with a twist tie. These jars and bags keep out oxygen and humidity, which stimulate microbial activity that deteriorates herbs. In addition, herbs hold their quality longer if stored away from any source of light (natural or electric) in a cool cupboard or closet. Exposure to light and/or heat fades their color and weakens their medicinal qualities.

Dried leaves and flowers are best kept up to one year, though you may find that some of your herbs hold their color, smell, and taste for longer. I like to refresh my herbal cabinet each autumn with the herbs I gathered in spring and summer. Some of the older herbs, such as comfrey leaves, holy basil, lemon verbena, calendula, and sage, I save for making into footbaths or baths for clearing away stressful or disharmonious energy.

Dried roots and barks can be stored for approximately two years.

Building a Rodent-Proof Drying Room

At Avena we rodent-proofed our drying rooms by applying quarter-inch wire mesh to the walls, floors, and ceilings, then covering the insides of the walls and ceilings with Sheetrock and the floors with pine boards. One of the rooms has insulation between the wire mesh and the Sheetrock. (If you live in a northern climate, I recommend insulating your drying room.) If you are building or retrofitting a space for herb drying, look into nontoxic Marmoleum for flooring, as it is much easier to clean than wooden flooring and there are no cracks for pantry moths to lodge their eggs in. If you are a small farmer seeking money to

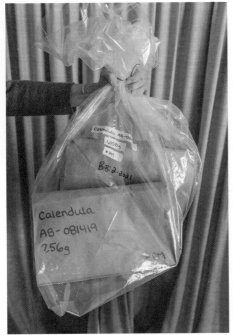

ABOVE
Rodent-proof wire-mesh screening surrounds the outside walls, floor, and ceiling of Avena's drying rooms.

RIGHT
Dried and weighed calendula flowers, stored in brown paper bags and then in plastic bags, waiting to be used in teas, salves, and other products

build a drying room, explore funding from your state's small-farming organizations.

Coping with Pantry Moths

Pantry moths often appear when comfrey leaves come into the drying room. These moths can multiply quickly, and their larvae will destroy your bagged and stored herbs. If you are concerned that there may be eggs on your herbs, place the dried herbs into a clean brown paper bag and roll shut. Then place the paper bag inside a sturdy plastic bag, seal it with a twist tie, and leave it in a freezer for three to ten days. We place Safer brand nontoxic pantry moth traps in various spots around the floor of Avena's herb drying rooms and keep a pantry moth notebook to record the date and number of moths found on each trap, renewing the traps as needed.

Tools for the Drying Room

- **Yearly plant-receiving logbook and pen.** Record the date, weight, and common name of each fresh herb brought into the drying room, as well as the weight when dried and date when bagged. Over the years, this information has been extremely useful in understanding the fresh-to-dry ratio of each herb we grow and the approximate drying time for each specific flower, leaf, and root.
- **Sharpie pens and masking tape.** Each time we lay an herb onto a drying screen, we write the name of the herb and the date on masking tape and stick the tape to the wooden frame. The date becomes part of the lot number we use in

our record-keeping system. For example, in lot number AB09012019, the AB stands for Avena Botanicals and the date is September 1, 2019.

ᴡ **Calculator.**

ᴡ **Good-quality scale** for weighing herbs. If you are selling dried herbs, you will need to calibrate your scale annually and record the date and method used.

ᴡ **Medium-size herb basket** for weighing fresh and dried herbs.

ᴡ **Thermometer and humidity gauges and daily heat and humidity log.**

ᴡ **Drying screens and racks** (see above).

ᴡ **Designated gloves** for stripping dried leaves and flowers from stalks.

ᴡ **Designated gloves** to wear when cleaning the drying rooms.

ᴡ **Nose and mouth mask.**

ᴡ **Designated drying room broom and dustpan.**

ᴡ **Designated drying room vacuum and vacuum bags.** We store our vacuum outside the drying rooms. If you are drying herbs for yourself and not for sale, you can be more creative with your tools for cleaning the drying space.

ᴡ **Designated brush for cleaning screens and racks.**

ᴡ **SaniDate 5.0 spray**, a nontoxic cleaner approved by the FDA and our biodynamic and organic certifiers.

ᴡ **Safer brand nontoxic pantry moth traps** (see above).

ᴡ **Floor mats** for happy feet.

ᴡ **Brown paper bags** for dried herb storage. Use your Sharpie to write the name of the herb, the weight, and the lot number on each bag.

ᴡ **Compost bin.** We keep ours outside the drying room for plant material that comes from the floor or from cleaning the screens.

General rules for entering and working in the drying room:

- ❦ Wash your hands before collecting herbs and/or when entering the drying room.
- ❦ Take off your shoes.
- ❦ Keep long hair covered with a bandanna, hat, or hairnet, whether you're drying herbs for yourself or selling the dried herbs.
- ❦ In a designated notebook, label and record the date and weight of all plants moving into and out of the drying room. Accurate records are helpful for reference year after year, no matter the size of your garden or farm.
- ❦ Cover your mouth and nose with a mask when stripping dried herbs to prevent inhaling fine particles of plant material and dust.
- ❦ Drink plenty of water.
- ❦ Keep the door closed at all times to prevent a rodent, cat, or dog from entering.
- ❦ Handle screens and plants with care and thoughtfulness. We deep-clean our drying screens and rooms every spring, and carefully brush them off by hand or with a designated brush throughout the season.
- ❦ Keep your thoughts and conversations kind and considerate when working in the drying room. Consider listening to uplifting music or podcasts when you need inspiration to keep you in the present.

Options for Drying Herbs

Gas Stove

A gas stove with a pilot light in the oven can be a useful space for drying herbs, although in today's world of modern kitchen appliances, gas stoves with non-electric pilot lights are less common. Use an oven

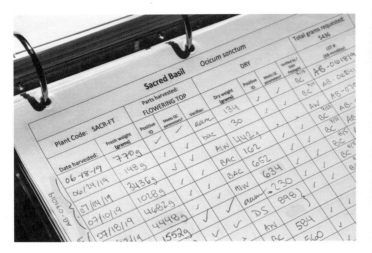

thermometer; the ideal temperature to dry herbs is 75–100°F (24–38°C). You can make simple herb-drying racks with screens that slide on top of your oven racks. Crack the oven door slightly and tape a note to your oven door to remind yourself that herbs are in the oven.

Electric Dehydrators

My friend, the longtime biodynamic herb grower Jean-David Derreumaux, uses a commercial-grade electric food dehydrator from Cabela's for drying medicinal flowers such as chamomile and calendula.

Greenhouses

Two wonderful organic herb farms in the Northeast, Zack Woods Herb Farm and Healing Spirits, dry hundreds of pounds of high-quality herbs in their greenhouses. In their book *The Organic Medicinal Herb Farmer*, Jeff Carpenter and Melanie Carpenter of Zack Woods provide excellent information on growing and drying herbs in greenhouses and drying sheds.

The colors of high summer: freshly harvested calendula, lemon bergamot, goldenrod, *Schisandra* berries, lemon verbena, and nettle

OVERLEAF
Lavender patch with mullein and black cohosh near one of Avena's oak trees

Beauty is not peripheral, but at the core of what sustains us. Awe and wonder ignite our imagination. We are inspired. We witness the magnificent and miraculous nature of creation—where there is harmony there is wholeness. The world is inter-connected and interrelated.

—Terry Tempest Williams, *Erosion*

Part III

PREPARING HERBAL MEDICINES

Herbs benefit our health in many ways. When taken regularly, nutritive teas, culinary herbs, and herbal soup broths add minerals and vitamins to our diet, support digestion, and enliven our food with their fragrances and flavors. Tinctures, glycerites, syrups, vinegars, and oxymels (made with various solvents, such as food-grade alcohol, organic vegetable glycerin, raw honey, or apple cider vinegar) address a wide range of health needs, from relieving acute illness to assisting in long-term well-being. Herbal steams aid sinus, respiratory, skin, and emotional health. Footbaths relax the nervous system, ease anxiety, improve sleep, and support various musculoskeletal conditions. Herbal oils and salves soothe cuts and wounds, inflamed joints and tendons, sprains and strains, sore muscles, and bruises. Hydrosols, also known as "floral waters," are used topically as an uplifting aromatherapy spray or as an ingredient in body-care creams and lotions. Flower essences transform limiting emotions, behaviors, and attitudes into more heart-centered and life-affirming ways of living.

Tenderly touching, tasting, smelling, and ingesting plants connects you with their physical healing qualities and spiritual essence. Let the plants inspire your imagination and intuition when preparing herbal medicine. Open your heart to the spirit and magic herbs offer. Invite a friend to join you, especially if you are new to herbs. Keep good notes. Build upon your previous experiences. I agree with herbalist Nathaniel Hughes's words:

Though the practical fundamentals of medicine making are easy to grasp, creating vibrant medicine from living plants can be a lifetime's vocation. A truly potent medicine has within it the essence of the plant; it is the herbalist's work to learn how to invite this essence out of the plant and into the bottle.

PREVIOUS
A ruby-throated hummingbird in Avena's garden enjoying nectar from the native jewelweed (*Impatiens capensis*) flowers

OPPOSITE
Freshly harvested rose petals from Avena's biodynamic *Rosa rugosa* hedgerow

Growing and preparing herbal medicine is an ancient art and a human right, and I hope that someday soon everyone, everywhere, will have the freedom, support, and skills to engage in this healing practice. Let laughter, love, gratitude, and humility infuse your heart and hands and the medicine you create.

Herbal Tea

Preparing herbal tea is an age-old ritual, an act connecting us with the elements—earth, water, air, and fire—and the wisdom radiating from the cosmos—the sun, moon, planets, and stars. Herbal teas can be made with roots, leaves, buds, flowers, seeds, berries, or barks. Whether you prepare them from fresh or dried herbs depends on what's available and whether a specific herb's healing qualities are more potent when the plant is fresh or dried.

The flavor of a healing tea is always a consideration, whether it is a medicinal tea or simply a cup of mint tea to share with a friend. An herbal tea made for medicinal purposes is usually stronger in taste than a tea primarily made for pleasure. Larger quantities of herbs are involved, and the medicinal tea may infuse or decoct (simmer and concentrate) for a longer time. The herbs chosen are specific to the healing process desired. Learning the flavor and healing properties of individual herbs is the first step in understanding how to combine them wisely. Knowing their flavors and ideal steeping time develops with experience and curiosity.

Herbal Infusions with Flowers and Leaves

A tea made by steeping or infusing the more delicate parts of plants—leaves, flowers, and aromatic seeds (for instance fennel, anise, or

LEFT
Preparing herbal tea
with fresh mint and
Tulsi

BELOW
Dried lemon verbena for
use in teas or honeys

Lemon Verbena

dill)—in hot steaming water is called an infusion. When making an infusion from dried herbs, gently crumble the leaves into a glass quart jar or tea strainer before adding hot water. I prefer to leave flowers whole so I can see them unfurl in the hot water.

Storing dried leaves and flowers as whole as possible preserves their flavor and medicinal properties; once an herb is crushed or powdered, the oxidation process begins and the herb's potency starts to diminish. Kept in a glass jar in a dark cupboard, away from heat and light, most dried leaves and flowers have a shelf life of twelve to fifteen months. Some dried herbs, like lemon balm, lose their flavor more quickly than others. I prefer dried lemon verbena and lemongrass as tea in winter, as they hold their flavor longer.

LEFT
Pouring a cup of freshly
brewed nettle tea

ABOVE
Dried nettle leaf waiting
to be used in teas

Herbal Infusion Made from Dried Herbs

1. To make one quart (about one liter) of tea, place 4–6 tbsp. (59–89 ml) of dried herbs into a glass quart jar, 32 oz. French press, or teapot. For a single cup of tea (approximately 8 oz. [237 ml]), place 1–2 tbsp. (15–30 ml) of dried herbs into a tea strainer.
2. Fill the jar, teapot, or mug with hot, steaming water and cover with a lid or small plate. Infuse for ten to thirty minutes.
3. Strain the herbs and place them in your compost. If you do not have access to a compost pile, consider putting the spent herbs in houseplants or outside under trees or bushes.
4. Enjoy holding a cup of warm tea in your hands, smelling the aroma, and imagining the spirits of the herbs infusing you with their wisdom and healing gifts.

To make an extra-strong medicinal infusion, cover 1 oz. by weight (28 g) of dried herbs with one quart of hot, steaming water. Cover and let steep four to eight hours or overnight. Strain and squeeze the herbs through cheesecloth (unbleached or organic—wash and reuse) to extract more of the liquid. If you do not drink the quart of tea in one day, store the leftover tea in a refrigerator or cool place and drink it the following day.

Herbal Infusion Made from Fresh Herbs
When I was a teenager learning how to make herbal teas, I intuitively sensed that fresh, living plants did not enjoy having boiling water poured over them. Hot water seemed appropriate for dried herbs, but not for fresh ones. Since that time, I have always placed handfuls of

fresh herbs into a glass or stainless cooking pot, covered them with cool water and a lid, and slowly heated the water to a simmer as a gesture of respect for the plants' sentient nature and as a way to preserve the delicate aromatic oils.

Fresh herbs embody a life force, a vital energy, that I don't sense as strongly when drinking dried herbal tea. Explore this for yourself. My favorite fresh herbal teas include holy basil, lavender blossoms, lemon balm, lemon verbena, rosemary, mountain mint, borage blossoms, calendula flowers, heartsease pansy flowers, red clover blossoms, marshmallow leaves, mullein flowers, roses, and green milky oat seeds. All but mullein and heartsease pansy flowers can be dried for winter teas.

LEFT
Fresh borage, calendula, roses, comfrey, and goldenrod

RIGHT
Sun tea made of fresh rose, mint, and lemon balm in spring water—ready to drink after sitting in the sun for two to three hours

1 Whenever possible, let the gathering of fresh herbs be a healing time. Even if for just five minutes, slow down and gather the herbs mindfully. Place handfuls of fresh, whole herbs (without shredding or cutting them) into a glass, enamel, or stainless cooking pot. Enjoy touching, smelling, and tasting these wonderful herbs.

2 Completely cover the herbs with cool water and a lid. Slowly heat to a simmer but not boiling. Stay near the stove; it is easy to forget to turn off the heat if you leave the kitchen. Once you turn off the heat, let the tea infuse as long as you wish—ten to thirty minutes, or even all day.

3 You may strain the tea through a colander or stainless strainer, then compost the herbs. I often pour both the herbs and the hot water into a quart jar and sip this tea all morning—I enjoy seeing flowers and leaves floating in my tea.

A Gardener's Sun Tea

Creating a sun tea is one of my favorite ways to connect with the garden early in the morning. Sometimes, I quietly walk in the garden and gather the plants that call out to me. Other times, I already know exactly which herbs I need. Though a sun tea will be infused with fewer chemical constituents and less flavor than a tea made from herbs infused in hot water, the sun's energy adds an extra sparkle and soulful warmth.

Depending on how many people you are serving, fill a quart, half-gallon, or gallon (one to four liter) glass jar with fresh herbs. Completely cover with cool water and a secure lid. Place the jar in a sunny location for several hours. The quart jar I make in the morning sits near me as I work in the garden. By lunchtime I've finished it and I refill the jar, covering the same herbs with water once more.

Many of us take water for granted. Sipping sun tea invites me to pause and consider the preciousness of plants and water. How I hope that soon, everyone, everywhere will value and protect the watersheds, wells, ponds, lakes, rivers, bogs, marshes, and streams we depend upon and that everyone, everywhere—no exceptions—will have access to clean water.

Magical Lunar Infusions

Moon tea possesses subtle rhythmic and magical qualities. Place handfuls of fresh or dried herbs in a glass bowl or pot, cover them with water, and place in the garden or in an outdoor location where the moon's light will infuse the water with her magic. If nighttime insects are common in your area, cover the bowl with a clear glass lid or thin cotton cloth. Before sunrise, enjoy drinking this magical infusion and/or pouring it over your hands and face.

Spirit Infusions

Rocio Alarcón Gallegos, a gifted healer, teacher, and ethnobotanist from Ecuador, has expanded my awareness of plants and herbal infusions. Herbalists tend to think of infusions as teas only, but Rocio has guided me to recognize that when we spend time in the presence of a particular plant, watch a hummingbird, or sit quietly in the garden, an infusion process is also occurring. By consciously paying attention, I invite the healing spirit of a plant, bird, or place to infuse my energy field. These experiences evoke inner transformations, opening the heart and mind and expanding compassion and consciousness. Rocio says that transformations help us "put our hands into action for the Earth."

A hummingbird rests on a magnolia tree, surveying which flowers to visit next. Avena has many flowering plants that ruby-throated hummingbirds love, including red bee balm, hummingbird sage, colorful nasturtiums, and tiny white nicotiana flowers.

Herbal Decoctions with Roots, Barks, Seeds, and Berries

A decoction is made by simmering the woody parts of plants (roots, twigs, and tree barks) or seeds and berries in water for twenty to forty minutes (or longer for specific roots, seaweed, and mushrooms). Simmering works on the cellular structure of the more fibrous plant parts; with the aid of heat and time, they impart their medicine to the water.

1 Place 5–6 tbsp. (74–89 ml) dried herb pieces or 8–10 tbsp. (118–148 ml) freshly chopped roots or shredded bark into a glass, enamel, or stainless cooking pan and cover with a quart of cool water and a lid. You can let these herbs sit overnight and then slowly bring them to a simmer for thirty minutes, or simmer them right away for thirty to forty-five minutes.

When I prepare teas from seeds, I use about 1 cup (240 ml) of water and 2 tbsp. (30 ml) of seeds. For the more delicate aromatic seeds, such as anise, coriander, cumin, dill, fennel, and freshly grated ginger, cover with cool water and slowly warm to a simmer, turn off the heat, and infuse, covered, for five to fifteen minutes.

2 Strain, and compost the herbs or seeds. If you have access to refrigeration, you can store leftover tea for up to two days. If you don't (like me), make your tea on a daily basis.

I enjoy drinking many medicinal roots as tea, but there is one whose flavor is too strong for me: valerian. I prefer to take valerian as a fresh root tincture. If you do make valerian root tea, infuse the dried roots in hot (not boiling) water for ten to thirty minutes or cover the fresh roots with cool water, heat the water until steaming, and steep, covered. The

I often use this old stainless cooking pot to make bark and root decoctions. Ingredients in this tea include cinnamon bark, astragalus, *Codonopsis*, and burdock and licorice roots.

volatile oils of valerian root are delicate, and best preserved by infusing in hot water rather than simmering.

General Guidelines and Safety Considerations When Using Herbs

Dosages vary depending on the person, the herb, the type of preparation, the health condition, and the herbalist.

❧ If you are new to herbs, start with small doses—a few drops of a tincture or glycerite or a few sips of tea—and tune in to how you feel physically, emotionally, mentally, spiritually. Do you feel a resonance? Do you feel agitated? Gradually increase the dose if needed.

❧ If you feel overstimulated, your health condition worsens, or you have any negative reactions, stop taking the herbs and seek professional advice from an experienced herbalist or herb-friendly medical provider.

❧ If your sleep becomes disturbed, either lower your dose, take the herbs before early afternoon, or consider changing the herbs.

❧ If you are pregnant, nursing, or using pharmaceutical medications, seek guidance from a professional health care provider who is knowledgeable in herbal medicine.

❧ For acute conditions such as a fever, flu, or earache, you should see a reduction in symptoms within the first twenty-four hours; if not, seek medical attention.

❧ When working with tonic herbs for strengthening body systems or supporting a chronic condition, you may need to increase your dose, use different herbs, make dietary changes, or investigate additional therapies if you don't sense any shift within the first four to six weeks.

Looking through the many jars of tinctures stored in a special room in Avena Botanicals' apothecary

ABOVE
Smelling a freshly
harvested rose

RIGHT
Some of my most
trusted texts

Acute Conditions

Tinctures, glycerites, syrups, and vinegars are helpful to have in your medicine cabinet for acute conditions commonly experienced in your family, neighborhood, or workplace. Treating acute conditions like a flu, fever, earache, or menstrual cramps requires the frequent use of herbs that address the specific symptoms. Timing depends on the person and the condition: herbs can be taken every fifteen to twenty minutes for the acute pain of menstrual cramps or intestinal gas, or every one to two hours at the onset of a flu or sore throat. Once the acute symptoms subside, dosing can be reduced to three to four times a day. Additional or even different herbs may be called upon to ease constriction and emotional sensitivity and to support overall digestion.

Chronic Conditions

Tonic herbs chosen to address chronic conditions vary widely and depend on each person's overall health and history: age; constitution; life experiences, including trauma; access to food; use of pharmaceuticals; and any other healing modalities that are being employed. The recommended dose is one to three times a day, evaluating the effectiveness every three months.

When addressing chronic health issues with nourishing teas and herbal tonics, it is beneficial to blend in herbs that ease stress, calm the mind, and support digestion and immunity. Flower essences, used together with a nourishing tea and tonic tincture, help release trauma and shift old patterns, support somatic mindfulness, and enhance heart-centered and peaceful ways of being. When working with preverbal or other serious traumas, seek out a professional practitioner—if possible, one specifically trained to work with trauma. Lack of access to herbalists and herbs, trauma-trained professionals, and good-quality food are some of the barriers facing many folks seeking health care.

Herbal Tonics

Herbal tonics strengthen and support the health of specific organs and body systems, and improve overall physical stamina and emotional and mental well-being. Use them one to three times daily and for extended periods, like one to three months or longer. Certain tonics are appropriate during different seasons of the year or different seasons of one's life, such as the menstruating, menopause, or elder years. They are useful following an injury like a broken bone, or for strengthening the lungs after exposure to a forest fire or serious virus. Tonic herbs can be used as teas, tinctures, glycerites, syrups, powders, and in foods such as hot cereal, smoothies, nut butter balls, and soup broths.

Herbal Tinctures

Tinctures are concentrated liquid extracts made with a food-grade alcohol. They are easy to make and have a much longer shelf life than dried herbs. Tinctures are helpful for treating various first-aid or acute conditions, for travel, and as donations for use at community clinics. Alcohol works well as a medium for herbs that you prefer taking as a tincture instead of a tea (goldenseal, valerian root, boneset) and for herbs that do not extract their medicinal properties into water (poplar buds, bee propolis). If you live where fresh herbs are not available year-round, tinctures enable access to specific herbal medicine when winter lingers. Alcohol tinctures are not recommended for children under the age of two, people addicted to alcohol, or anyone who refrains from alcohol for health or religious reasons.

An alcohol tincture is quickly absorbed into the body, as it bypasses digestion and directly enters the bloodstream. The amount of alcohol in an individual tincture is based on the alcohol percentage used to make the tincture. For example, a dropper of tincture (made with 50 percent alcohol) contains less alcohol than the amount of naturally occurring alcohol in a glass of commercial orange juice or an overly ripe banana. The body, via the liver, has a pathway to detoxify alcohol and turn it into a sugar. The shelf life of most herbal tinctures when made well and stored in a cool, dark cupboard ranges from four to eight years.

We prepare most of the tinctures in Avena Botanicals' apothecary with fresh herbs from the garden. Whether you prepare your tincture from fresh or dried herbs depends on your personal preference, which herbs you have available, and the unique biochemical and energetic qualities of each herb. There is no single right way to make tinctures. Record your recipes and experiences in a notebook and talk with other herbalists.

Alcohol Percentages

A 100 proof (50 percent alcohol) vodka (organic when possible) works well for most dried and fresh plants. When making tinctures from a resinous substance like myrrh, bee propolis, or *Boswellia*, a higher alcohol percentage is recommended (150 to 190 proof). If you use 190 proof alcohol for all your tincture making, you will need to dilute it with water when you need a lower alcohol percentage. For example, mix equal parts 190 proof alcohol and filtered or spring water to get the desired 100 proof (50 percent) alcohol for your menstruum (the liquid part of your tincture).

LEFT
Grinding fresh nettle
into tincture

ABOVE
Freshly ground nettle
tincture

Dosage for Herbal Tinctures

A tincture's dose and timing may depend on the type of herb; your age and weight; your alcohol sensitivity; and whether your condition is acute or chronic. Best to start with smaller doses (three to five drops) if you are new to taking a specific herb, and see how your body feels before increasing the dose. My thirty-five years of experience as an herbalist and herbal medicine maker have led me to believe that the standard dose the FDA requires tincture makers to put on our labels is too high. Start with a lower dose, and if possible (especially when working with an acute condition) stay in communication with an herbalist. (We need herbalists rooted in every community and integrated into the public health care system!) Herbs are not meant to overstimulate but to balance, strengthen, and heal us. So start with smaller doses and increase mindfully and skillfully.

The viscosity (thickness) of tinctures varies, depending on the specific herb used. Generally, one dropperful of tincture is around thirty to thirty-five drops or ¼ tsp. (1.2 ml). A 1 oz. bottle contains 30 ml.

Acute dose for adults: ¼–1 tsp. (1.2–5 ml), four to six times per day
Tonic dose for adults: ¼–1 tsp. (1.2–5 ml), one to three times per day

Acute dose for children, ages 3–10: one drop of tincture per 5 lb. (2.3 kg) of body weight, four to six times per day
Tonic dose for children, ages 3–10: one drop of tincture per 5 lb. (2.3 kg) of body weight, one to three times per day

I use several herbs, including teasel root, blue vervain, and rose petal elixir, in smaller doses (three to ten drops total). Some herbalists refer to small doses as Spirit drops. Let the plants and your intuition guide your use of Spirit drops for shifting emotional and mental patterns, opening and healing the emotional heart, and uplifting the spirit. My friend Kay Parent, a wise and loving clinical herbalist, gives her clients a beeswax candle to light each time they take their herbs, inviting them to slow down and be present and mindful as they ingest their plant medicines.

Preparing a Fresh Plant Tincture

1. After carefully collecting the plants, check them and compost damaged parts, such as rotten root sections or yellow or chewed leaves.
2. Wash the roots. Leaves and flowers generally are not muddy like roots and should not be washed.
3. Choose a glass jar that will accommodate the volume of herbs you gathered. I tend to place whole flowers in the jar, leaving a few inches of headroom, and then fill the jar to the top with a 100 proof (50 percent) organic alcohol. For leaves and roots, either chop them or grind them in a blender with the alcohol, then pour the mixture into a glass jar. Secure the lid tightly. I like to shake the bottle before placing it on the shelf—sometimes singing, sometimes saying prayers, sometimes just smiling and dancing around the room while gently shaking the jar.
4. Label and date the tincture. Include the name of the plant and the parts used. If you weighed the plant before putting the plant in the jar or blender and/or measured the alcohol, record the weight and liquid measurements in your apothecary notebook. Include the place you harvested the plant and any interesting weather information or bird observations.

Making an echinacea
flower tincture

5 Place the jar in a dark closet or cupboard and let it sit for four to eight weeks. Shake it regularly, at least a few times a week. My heart believes that singing, chanting, or carefully dancing when shaking a tincture adds healing energy. Enjoy watching the liquid take on a new color as the plants pass on their medicine. During the first week of extraction, you may need to top up your jar as the plants absorb liquid. This ensures that the plants stay completely covered in the liquid (menstruum) while tincturing.

6 After four to eight weeks, strain the tincture through unbleached or organic cheesecloth. An easy way to do this is to place a stainless colander in a large bowl, drape the cheesecloth in the colander, and pour the liquid in. Tightly wring the cheesecloth, which contains the plant matter, to release as much liquid as you can. Compost the plant material.

7 Pour all the liquid into a glass bottle with a tight-fitting lid. Label and date your tincture and store in a cool, dark place. Tinctures have a shelf life of approximately four to eight years. They will taste weak and smell funky if no longer viable.

Preparing a Dried Plant Tincture

1 Place 5 oz. (142 g) by weight of a coarsely chopped dried herb in a glass quart jar (for a pint, use 2 ½ oz. [715 g]). A scale that measures grams, ounces, and pounds is a helpful addition to an herbalist's kitchen.

2 Standard tinctures use a weight-to-volume ratio of 1:5, though I prefer a 1:3 or 1:4 ratio. So the amount of 100 proof alcohol needed for 5 oz. (142 g) of dried herbs, at a 1:4 ratio, is 20 oz. (600 ml).

Placing dried hawthorn berries into a jar for use in winter teas. These hawthorn berries have spent at least two weeks in the drying room, followed by two weeks in brown paper bags, before being moved to a glass jar.

Most glass measuring cups read in both milliliters and ounces, though they don't measure as exactly as a glass beaker.

3. Cover the herb with the measured 100 proof vodka. Pour into a blender and grind. If you don't use a blender, just put a tight lid on your jar and label it. Follow the remaining instructions above for making fresh plant tinctures.

Herbal Glycerites

Glycerin is a sweet, sugar-free, alcohol-free, mucilaginous liquid derived from vegetable sources: sugar beets, coconut, soy, flax seeds. When sourcing it, be sure to purchase a certified organic vegetable glycerin base (available online). This will ensure that it is not made from genetically modified organisms (GMOs) (sugar beets and soy are commonly genetically modified). Technically, glycerin is a triatomic alcohol, not a sugar, so in healthy folks it does not affect blood-sugar levels. Glycerin extracts aromatic oils from leaves and flowers and the mucilage found in roots like marshmallow and comfrey.

No one taught me how to make herbal glycerites from fresh herbs—I used my taste buds and some understanding of plant chemistry to create Avena Botanicals' herbal glycerites. I much prefer the flavors of holy basil, rose, lavender, lemon balm, peppermint, fennel seed, catmint, and ginger in a glycerin base rather than an alcohol base.

Preparing a rose petal elixir using *Rosa rugosa* petals, organic glycerin, and a small amount of organic alcohol. Traditionally herbal elixirs are made with honey.

At Avena we make elixirs with ¾ parts organic glycerin and ¼ part organic alcohol.

Glycerin extracts the valuable essential oils of these fragrant herbs really well, creating a delicious-tasting, non-alcoholic herbal medicine. A glycerite's shelf life is around two years.

Glycerin is not as strong a solvent as alcohol, so some people dose a bit higher with glycerites than with alcohol tinctures. Glycerites work well for children, alcoholics, and anyone sensitive to alcohol. They also work well when combined with tinctures, adding a sweet and pleasant flavor to herbal blends.

Preparing a Fresh or Dried Plant Glycerite

1. After carefully collecting the plants, check them and compost damaged parts, such as rotten root sections or yellow or chewed leaves.
2. Wash the roots. Leaves and flowers generally are not muddy like roots and should not be washed.

3. Choose a glass jar that will accommodate the volume of herbs you gathered. When using fresh flowers, I tend to place the flowers into the jar whole, leaving an inch of headroom, then fill the jar to the top with organic vegetable glycerin. (I do *not* add water when making glycerites from fresh or dried plants.) A general guideline when making fresh plant glycerites is to use 1 part herb to 2 to 3 parts organic glycerin (by weight). When making a dried plant glycerite, use 1 part herb to 4 to 5 parts organic glycerin (by weight). I recommend grinding leaves or roots with the glycerin in a blender; the grinding process creates heat, and this heat aids in the extraction of the plants' constituents. Pour the blended glycerite into a glass jar to macerate (steep), leaving a few inches of space, and secure the lid tightly.

4. Label and date the jar of glycerite or elixir. Include the name of the plant and the parts used. If you weighed the plant before putting the plant in the jar or blender and/or measured the glycerin, record the weight and liquid measurements in your apothecary notebook. Include the place you harvested the plant and any interesting weather information or bird observations.

5. Place jars in a dark closet or cupboard and let the glycerite macerate for four weeks. Shake your jars regularly, at least a few times a week. Enjoy watching the liquid take on a new color as the plants pass on their medicine. Be joyful and grateful when you shake your medicine.

6. After four weeks, strain the glycerin through unbleached or organic cheesecloth. An easy way to do this is to place a stainless colander in a large bowl, drape the cheesecloth in the colander, and pour the liquid in. Tightly wring the cheesecloth, which contains the plant

LEFT
Fresh *Rosa rugosa* petals recently placed in organic glycerin and alcohol.

RIGHT
After a few weeks, the rose petal elixir begins to take on the color of the petals.

matter, to release as much liquid as you can. Compost the plant material.

7. Pour all the liquid into a glass bottle with a tight lid. Label and date your herbal glycerite and store it in a cool, dark place or in the refrigerator. Glycerites have a shelf life of approximately two years if stored properly. They taste unpleasant when expired.

Herbal Vinegars

Raw, organic apple cider vinegar is my favorite kind to use when creating mineral-rich and culinary herbal vinegars, as it is full of beneficial bacteria and enzymes. Raw apple cider vinegar infused with mineral-rich herbs (such as fresh or dried alfalfa, nettle, oat straw, horsetail, fresh chickweed, parsley, purslane, cleavers, violet leaf, plantain leaf, lamb's-quarter, and green milky oat seed) helps strengthen and build bones. Unlike alcohol or glycerin, vinegar extracts minerals from the herbs, although it does not extract a wide range of medicinal constituents in the same way alcohol or glycerin does.

Infusing bitter herbs (blessed thistle, artichoke leaf, rosemary, burdock root, dandelion leaf and root, yellow dock root) in raw apple cider vinegar creates a supportive digestive blend to take before meals (1–3 tsp. [5–15 ml] diluted in a few ounces of water). Herbal vinegars add flavor and nutrients when blended into salad dressings or sprinkled on top of steamed greens such as kale, collard, or lamb's-quarter. Even without infused herbs, raw apple cider vinegar is already antiseptic, and improves digestive function and supports healthy gut flora. Start with ⅛–¼ tsp. (0.6–1.2 ml) diluted in 1 oz. (30 ml) of water. (However, if you have acid reflux, heartburn, or other gastrointestinal conditions, you may find diluted vinegar too irritating to the digestive tract.)

To create a colorful and flavorful culinary vinegar using herbs such as opal basil, nasturtium flowers, dill, calendula, or chive blossoms, try an organic distilled white vinegar, rice vinegar, or white wine vinegar.

Herbal vinegars are a wonderful addition to the kitchen medicine cupboard. They can be stored in a cool, dark place for one to three years. Especially in hotter climates, you may want to store vinegars and oxymels in a refrigerator (if one is available), to prolong their shelf life.

Preparing a Fresh or Dried Mineral-Rich Vinegar
If you are using fresh herbs, chop them with a knife or clippers. Gently fill a glass jar to the top with the herbs and completely cover them with an organic vinegar of your choice. Use a plastic lid; vinegar corrodes metal lids.

Rosemary freshly harvested from the garden can be a supportive addition to vinegars, teas, tinctures, footbaths, and healing baths.

Label and date each jar and place in a dark cupboard; sunlight can diminish the quality of the vinegar. Let the vinegar infuse for about a month, shaking regularly. Strain the vinegar through unbleached or organic cheesecloth placed in a stainless strainer and compost the herbs. Store in a cool, dark place.

Herbal Oxymels

Herbal oxymels are made with equal parts raw honey and raw apple cider vinegar. Fresh or dried herbs both work, though I prefer fresh ones. Let your imagination, taste buds, and herbal knowledge guide you. A few favorites of mine include fresh whole rose flowers as a single-herb oxymel, and a multi-herb oxymel made with fresh whole calendula flowers, fresh red bee balm flowers, and fresh lemon

Making a fresh calendula and lemon bergamot oxymel with organic apple cider vinegar and raw honey

OVERLEAF
Bob Harden, Avena's beekeeper, checking on the honeybees

bergamot flowers. Our friend Zoe from Canada once made us a lovely oxymel with fresh roses, red clover blossoms, and cleavers.

Follow the instructions for making herbal vinegars, but using half honey and half vinegar. The shelf life of oxymels varies; my fresh oxymels tend to keep for twelve to sixteen months when stored in a cool, dark place or in the refrigerator.

Herbal Vinegar Hair Rinse

Apple cider vinegar also lends itself to a lovely hair rinse. My favorite herbs include fresh or dried nettle, rosemary, chamomile, and lavender. Nettle and rosemary support healthy hair growth. Rosemary stimulates hair follicles and promotes healthy blood circulation. Chamomile helps soften the hair. Lavender stimulates mindful alertness and self-awareness, restores strength and vitality to the nervous system, and uplifts the spirit.

Keep a glass jar of this vinegar in your bathing room or outdoor shower. You can dilute it in 50 percent water or herbal tea (made with the same herbs) or gently massage the undiluted vinegar into your scalp and hair. Let it soak in for at least five minutes before rinsing it out. During the summer months, enjoy sitting outside for a while before rinsing out the vinegar.

Herbal Honeys

Honey is truly a magical sweet food, packed with beneficial enzymes and possessing healing properties that are soothing and antimicrobial. Honeybee workers (females who live not more than six weeks) may be the hardest-working beings on our planet. Please, as a way to honor honeybees and all pollinators, consider who you purchase honey (and

other food) from. A bow to all the ecologically and spiritually minded beekeepers and organic growers! Consider also your use of toxic cleaning supplies and other garden, yard, and roadside chemicals you may be spraying—they pollute our waterways. Science has abundantly demonstrated the vulnerability of honeybees and all pollinators in the highly toxic world we inhabit.

For more than twenty-five years I have made an herbal honey with fresh holy basil leaves and flowers, using raw honey from the same local beekeeper. I prefer raw honey (honey that has not been heated above 95°F [35°C]). However, the high water content found in fresh herbs like holy basil can cause raw honey to ferment. The solution is to turn your jar full of fresh herbs and honey every day, gently moving and mixing the honey and herbs. Strain the herbs from the honey after two to four weeks. I prefer the flavor of honey made from fresh herbs, though you can experiment with dried herbs such as lavender, rosemary, thyme, peppermint, and sage.

Herbal honeys are wonderful to add to tea, oxymels, and syrups, or to dribble on special cakes and pastries. Honey also soothes sore throats (it's great added to herbal throat sprays) and eases coughs. (See the herbal cough syrup recipe under Herbal Syrups, page 108.)

Preparing an Herbal Honey

1. Start with a clean, sterilized glass jar. Gently fill the jar with fresh herbs, maintaining air space between them, and cover completely with honey. If using dried herbs, use either ½ cup (120 ml) of the dried herb for a pint jar or 1 cup (240 ml) for a quart jar, and completely cover the herbs with honey. Mix with a clean stainless

Avena's honeybee hives

knife, making sure that the herbs remain fully covered. Add more honey if needed.

2. Cover the jar with a tight lid and label it with the date and herbs used. Store in a cool, dark place for two to four weeks, gently turning it regularly to avoid fermentation.

3. Strain the honey and herbs through a stainless mesh strainer. I like to put the herbs back into the glass jar and cover them with warm herbal tea, so I can honor and enjoy the last drops of honey that remain in the jar and on the herbs. Then I compost the herbs.

4. Pour the strained honey into a glass jar, cover with a tight lid, label, and store in a cool, dark cupboard. Herbal honeys will keep for at least one year. If you are concerned about fermentation, store your jar in a refrigerator.

A pint jar of infused honey, when strained, yields about ¾–1 cup (180–240 ml) of honey. A quart jar will yield 2–3 cups (480–720 ml) of strained honey.

Herbal Syrups

Herbal syrups are concentrated, sweetened herbal teas that keep in the refrigerator up to six months, or longer if preserved with an herbal tincture or brandy. They are fun to make with family members and friends and to give as gifts to those hesitant to make herbal teas for themselves. Syrups are valuable to have on hand for acute conditions such as a cough, cold, fever, flu, indigestion, or crankiness.

Preparing an Herbal Syrup

1. Place 2 cups (480 ml) fresh or 1 cup (240 ml) dried herbs into a glass or stainless saucepan and cover with 4 cups (1 liter) of water. Slowly bring the water to simmer. Simmer until the water volume has been reduced by half. I use a glass measuring cup to aid in accuracy.

Sumac Honey Syrup with Schisandra + Hawthorn Berry Tinctures

2. Strain the herbs and tea through a stainless strainer and compost the herbs. Sometimes I squeeze the herbs through cheesecloth to extract even more of the healing tea.

3. Let the concentrated tea cool to 95°F (35°C). For every 2 cups of concentrated tea, add 2 cups of honey and stir well to dissolve the honey. When adding tinctures to a syrup, I use less honey: 2 cups of tea, 1 cup honey, 1 cup tincture. Pour into a sterilized quart glass jar and label with the date.

4. Once the syrup is cool, you can store it in a refrigerator for up to six months. If any mold forms, just scrape it off and pour the syrup into a different sterilized glass jar.

5. Because I live in a small solar home with no refrigerator, I add herbal tinctures to my syrups and store them in a cool, dark cupboard through the winter. If I have any syrups left by spring, I store them in the farm refrigerator through the warmer months. Syrups preserved with an alcohol tincture or brandy can last twelve to sixteen months when stored in the refrigerator.

LEFT
Cutting white pine
needles and twigs with
Felco 310 clippers

ABOVE
Thyme seedlings
growing in Avena's
greenhouse. They are
transplanted outdoors
when eight weeks old.

White Pine and/or Thyme Cough Syrup

Pine is cleansing and revitalizing to the respiratory system; a pine needle and twig tea, syrup, or tincture reduces coughs and respiratory congestion and helps resolve sinusitis, bronchitis, and pneumonia. My friend Kate Gilday says being around pine is like drinking fresh air. If you don't live around white pine trees (*Pinus strobus*), you can substitute other herbs that soothe and resolve coughs, such as horehound, lavender, or mullein leaves. The inner bark scraped from small pine branches can be made into a decoction to treat deeply stubborn coughs and congestion (simmering for fifteen minutes releases the antiseptic resins). Take this decoction internally as a tea, or use it as a steam inhalation to draw out hardened, green mucus from deep within the lungs.

Thyme is a warming and pungent herb with many valuable medicinal qualities. If you live in a climate where garden thyme (*Thymus vulgaris*) grows easily, I highly recommend growing a bed with at least ten plants or growing some in pots. Fresh thyme tincture is one of my favorite warming winter remedies for easing coughs, colds, and flu; relieving a variety of digestive disturbances; treating fungal infections; and improving circulation. Thyme relaxes the respiratory tract muscles, and possesses antimicrobial constituents and a pungent, warming, antispasmodic nature that is valuable in resolving lung infections; easing bronchitis, asthma, and whooping cough; and expelling excess mucus from the upper and lower respiratory tract.

1. Clip 2 cups (480 ml) fresh white pine needles and small twigs into a saucepan and cover with 4 cups (1 liter) of water. Simmer until the water volume has been reduced by half, to 2 cups of tea.

② Strain the herbs and cool to 95°F (35°C). Stir in 1 cup (240 ml) of raw honey.

③ Once the syrup is completely cooled, add 1 cup (240 ml) of thyme tincture as a preservative. Label and store in a glass jar in the refrigerator or in a cool, dark cupboard; it will keep for twelve to sixteen months. Use 2 cups of honey if you are making this syrup without thyme tincture.

Herbal syrups can be added to medicinal teas to enhance their healing or taken directly by the spoonful. Syrups generally taste yummy because of the honey. Remember to use syrups without alcohol for children and anyone sensitive to alcohol.

Acute dose for adults: 1–2 tbsp. (15–30 ml), four to six times per day
Tonic dose for adults: 1–2 tbsp. (15–30 ml), one to two times per day

Acute dose for children, ages 3–10: 1–3 tsp. (5–15 ml), three to six times per day
Tonic dose for children, ages 3–10: 1–3 tsp. (5–15 ml), one to two times per day

Herb-Infused Oils and Salves

Herb-infused oils and salves made with unrefined organic olive oil are readily absorbed into the skin and do not go rancid as quickly as those made with other oils. Avoid mineral oil, as it is petroleum-based. The skin is the body's largest organ of assimilation and elimination; what we put on our skin is absorbed into the body.

Preparing Medicinal Oils with Fresh Herbs

1. Fill a clean, sterilized pint or quart glass jar with fresh herbs. Collect the herbs on a sunny morning after the dew has dried. The herbs we gather at Avena include whole mullein flowers, Saint-John's-wort flowering tops (ground in the blender with oil), rosemary, lemon balm (dry-wilted overnight and then ground in the blender with olive oil), peeled and chopped garlic cloves, and pine needles and twigs.

2. Slowly fill the jar to the top with oil, completely covering the herbs. Use a stainless knife to release any air bubbles; the herbs should be totally immersed to prevent mold from forming.

3. Cover the jar with clean, unbleached cheesecloth—not a tight-fitting lid—so the oil and herbs can breathe. The water in fresh herbs can lead to mold, as water and oil don't mix well. If mold develops, just scrape it off with a clean spoon and compost.

4. We place glass jars of herb-infused oils in an 80°F (27°C) oven with a pilot light for two weeks. The oven maintains an even temperature throughout the infusion process. If you are using an oven, be sure the temperature is not above 100°F (38°C) and post an "Herbs in Oven" sign so no one accidentally turns the oven on. Gently stir the herbs and oil regularly with a clean stainless knife or spoon or a rubber spatula.

5. After two weeks, pour the oil through clean cheesecloth into another sterilized glass jar to separate it from the plant matter. Compost the herbs. Cover the new jar with cheesecloth and let it sit undisturbed for one or two days. The water and sludge from

ABOVE
High-quality Saint-John's-wort oil, made from fresh flowering tops and organic olive oil, infused at 100°F (38°C) for two weeks, becomes a beautiful deep red color.

RIGHT
Pine oil with cheesecloth

the fresh plants that mixed with the oil during the infusion process will settle at the bottom of the jar.

6. With a small stainless ladle, scoop off the oil sitting on top of the water and place it in another clean glass jar. This method prevents your fresh herb oils from fermenting. Once the oil is free of water, place a tight lid on the jar, label, and date, and store in a cool, dark place for twelve to eighteen months.

Preparing Medicinal Oils with Dried Herbs

Consider using these dried herbs: comfrey leaf and root, echinacea root, Solomon's seal root, calendula flowers, lavender blossoms and leaves, holy basil leaves and flowers, mugwort leaves and flowers, witch hazel leaves, yarrow leaves and flowers, rosemary.

1. Place 8 oz. (237 ml) of dried herbs and 8–12 oz. (237–355 ml) of organic olive oil into a blender and grind.

2. Pour the mixture into the clean glass pint jar. Cover with a lid, label, and date. Place in an oven warmed by a pilot light, in a sunny window, or in a sandbox for two to three weeks. Herbalist Juliette de Bairacli Levy lived in the Mediterranean region most of her life, without electricity, and Rosemary Gladstar tells the story that Juliette would place her jars of herbal oils into sand, warmed by the sun, to infuse.

3. At the end of the infusing period, pour through clean, unbleached cheesecloth, squeezing as much oil as you can from the herbs.

4. Store the oil in a glass jar with a tight lid in a cool, dark place. Label and date. Oils made from dried herbs have a shelf life of twelve to eighteen months.

LEFT
Calendula tincture
(fresh calendula flowers
ground in organic
alcohol) and dried
calendula flowers
infusing in organic
olive oil

RIGHT
Calendula flowers on
drying screens. Once
dried, they will be made
into calendula oil and
heal-all salve.

Medicinal Oils Made into Salve

Depending on the herbs they contain, salves heal cuts, abrasions, cracked and dry skin, sore nostrils, chapped lips, hemorrhoids, inflamed and tender vaginal tissue, swollen perineum and labia after giving birth, cracked nipples, diaper rashes, and more. Salves are easy, fun, and magical to make from herb-infused oils. Always remember to keep good notes. The first one I created for Avena Botanicals in 1985 is called Heal-All salve. The recipe has stayed the same all these years:

1. To make 10 oz. (296 ml) of salve, place 8 oz. (237 ml) of herb-infused oil into a glass or enamel pot or double boiler. Add 2 oz. (59 g) of beeswax to the oil and slowly warm on the stove, stirring occasionally. Beeswax melts at 140°F (60°C). Be sure not to over-heat or the melted beeswax will start to smoke.

2. Remove from heat and immediately pour the warm liquid into small glass jars or tins. I like to put a few drops of flower essences into each container before filling. Once the salve has cooled and solidified, put on the lids, label, and store in a cool, dark cupboard. Salves last for two years.

LEFT
Pouring warm calendula
salve into tins. The
salve will harden once
it cools.

RIGHT
Waiting for the freshly
poured calendula
salve to harden in the
tins before placing on
the lids

Connect with a local or regional beekeeper for high-quality beeswax.
Store it in a dry container. You can wrap a block of beeswax in a clean,
thick cotton towel and hit it with a hammer on a solid counter to create
smaller pieces. Use a scale to weigh the beeswax. If you don't have
a scale, take a 16 oz. glass measuring cup and pour in 8 oz. (237 ml)
of oil, then keep adding small pieces of beeswax until the liquid level
reaches 10 oz. (296 ml). Consider making salves with or for children,
elders, or folks living in shelters or prisons.

Freshly harvested
calendula flowers

Grow a Row of Calendula

Violence against women is prevalent worldwide. When I first learned
about the playwright Eve Ensler's work in the Congo region to assist
women who have been raped and violated (see vday.org), I wept and asked
myself, "How can I be of help?" The answer was to grow a row of calendula.
This grassroots initiative encourages individuals with any size garden to
grow a row of calendula, collect the flowers regularly, dry them, and make
salve. Contact your local or county organization that serves women and
children who have been raped, sexually abused, or sex trafficked, and ask
if you can donate calendula salves. Print out and give information to the
staff about the Grow a Row Project and about calendula's healing qualities
(see helpgrowarow.org).

My hope is that this initiative will inspire people everywhere to grow
and dry calendula and that it will create more dialogue, community support, and understanding within families, schools, the health care system,
and government agencies of the root causes of violence and oppression.
Planting seeds and sharing herbs is a way to bring about positive change.

Baths

Foot, hand, full body, and sitz baths are healing and rejuvenating to the body and spirit. Baths stimulate circulation to congested, cold, or crampy areas; move stagnant blood; relax and calm the nervous system; warm the body; break a fever; detoxify; relieve aches and pains; and reduce swollen perineum tissue.

Start by making one or two quarts of tea using various fresh or dried herbs. Let the herbs infuse in hot water for fifteen minutes. Strain, then pour the tea into a bathtub filled with hot water or a pan designated for a foot, hand, or sitz bath.

Footbaths

When making a footbath in winter, I enjoy putting a few large handfuls of dried aromatic and colorful herbs (calendula flowers, holy basil, lemon verbena, comfrey leaves) directly into my foot basin and pouring hot steaming water over the herbs, half filling the basin. I submerge my feet into this magical herbal bath when the temperature has cooled but still feels pleasantly hot. I keep a kettle of hot water next to my foot basin for adding additional heat as needed, as I like soaking for ten to fifteen minutes. Once finished, I wipe my feet dry with a towel and rub an herbal oil or salve into my feet. Usually, the next morning I dump the herbs and water outside under a tree. If you live in an urban area, consider composting the herbs and watering houseplants with the bath water.

Fresh flower footbaths are nourishing after a long day of gardening. I gather whatever flowers and aromatic herbs call to me. Calendula, holy basil, roses, lamb's-ear leaves, lavender, lemon balm, lemon verbena, comfrey leaves, sage, rosemary, mints, mullein leaves, zinnias,

Footbath made with calendula, goldenrod, roses, borage, comfrey, and nettle

and peonies are favorites. I place the fresh herbs in a stainless pot, cover with cool water, and slowly heat until the water is steaming and fragrant. I then pour the herbs and water into my foot basin and submerge my feet once the water is a comfortable temperature, breathing in the relaxing, healing aroma of the flowers and herbs.

One or two drops of pure essential oils such as lavender, rosemary, lemongrass, geranium, pine, sage, or ginger can be used in a footbath, or around ten drops for a full-body bath. Rosemary essential oil in a foot or full-body bath is my favorite for softening emotional edges, uplifting the spirit, and supporting a more wakeful and attentive mind. When purchasing essential oils, please choose a company that sells only pure (not synthetic) essential oils made from plants that are not endangered and grow easily, in great abundance.

Healing Flower Baths

Rocio Alarcón Gallegos, the aforementioned Ecuadorian healer and ethnobotanist, has generously taught many people the importance of regular aromatic herb and flower baths for clearing negativity, fear, anger, grief, confusion, and chaos from the body's energy field. Please refer to Rocio to learn this practice in greater depth, as it comes from her culture, not mine (see iamoe.org); I will share what she has given me permission to write. Rocio encourages all of us to clean our energy fields regularly by pouring aromatic herbal teas over our bodies at least once a week, using fresh or dried herbs. She instructs us to ask the plants for their help, call in their spirit, and pray for healing as we gather and make the tea, then pour the healing water over ourselves either outdoors or in an indoor shower. Create quiet space afterward to absorb the plants' healing gifts. I feel less

A copper vessel is used to scoop out water for a healing bath of calendula, holy basil, cosmos, and comfrey.

Herbal steam

emotionally reactive and more relaxed, calm, and heart-centered when I engage in this practice weekly.

Herbal Steams

Herbal steams are hot herbal teas made with fresh or dried herbs or hot steaming water containing a few drops of pure essential oils. Place a thick cotton towel over your head and around the steaming pot of herbs. Sit or stand comfortably with your head at a safe distance from the steam to avoid burning your nasal passages and skin. Inhale the healing oils through your nose or mouth.

Aromatic demulcent (soothing) herbs have many healing virtues, including clearing and calming the mind and nervous system; decongesting the sinuses, nasal passageways, and lungs; easing mucus membrane inflammation; and resolving upper and lower respiratory infections. Aromatic teas or essential oils to consider include lavender, chamomile, pine, rosemary, and holy basil. Soothing herbal teas include plantain leaf, marshmallow leaf, mullein leaf, horehound, chamomile, and lavender.

If you use a woodstove in winter, keep a pot of aromatic herbal tea steaming on it throughout the day, especially when you or a family member is ill or during a pandemic. This enlivens the winter house with plant fragrances and keeps the air clean and refreshed. If friends or family members choose to stay at home in their dying process, consider keeping an aromatherapy infuser with soothing fragrances in their room and elsewhere in the house, where it will permeate the air. Be sure to continue this process throughout the home for several days once the loved one has passed.

Flower Essences

Every flower contains a unique vibrational pattern. To create a flower essence that embodies this vibrational pattern, sun-infuse fresh flowers in spring water for a few hours or throughout the night during a full moon. The life force expressed within a particular flower essence helps to awaken regenerative qualities within the human soul. Flower essences support our inner capacity to transform limiting emotions, behaviors, and attitudes into more life-affirming and open-hearted ways of being and living in the world. Usually a flower essence is taken orally under the tongue, two to four drops, several times a day. Flower essences can be supportive for animals, too: place drops in their water dish, let them lick drops off your hand, or place the drops directly on their head.

LEFT
A flower essence being prepared from fresh zinnia flowers, encouraging playfulness and lightheartedness

ABOVE LEFT
Mexican sunflower (*Tithonia rotundifolia*), beloved by hummingbirds, monarch butterflies, and bees. As a flower essence, this brilliantly orange, velvety flower inspires joy and creativity.

ABOVE RIGHT
Daylily flower essence supports us in living more fully in the present moment.

Preparing a Flower Essence

1. On a sunny morning, slowly approach the flower you wish to make an essence from. Spend time sitting or standing quietly there, attuning with the flower's spirit. Let the flower know why you have come. Ask permission. If permission is granted, make an offering at the base of the plant.

2. Carefully hold a small, sterilized glass bowl filled with spring water under the flowers you are gathering. Gently clip the flowers into the water without touching them with your fingers. Fill the bowl with flowers and then place it near the flowering plants. Be careful to keep your own shadow from crossing the bowl of flowers.

3. The sun's (or moon's) energy will infuse the water with the flowers' energy. Let your intuition guide you as to the duration of this infusion—generally a few hours or more.

4. When the infusion process feels complete, gently remove the flowers from the water using the same clippers as tongs, and place these flowers near the plant you harvested them from. Offer some of the infused water to the plant.

5. Half-fill a 1, 2, or 4 oz. (30, 60, or 120 ml) sterilized amber-colored dropper bottle with the flower essence (a small stainless funnel works well). Finish filling the bottle with oak-aged brandy (for preservation purposes) and seal with a glass dropper. This bottle is known as the Mother Essence. Label, date, and store in a special basket or in your medicine cupboard. Mother Essences can last for several years if stored away from light, heat, and digital devices. Any extra flower essence water can be sipped, shared with others, or given back to the flowering plant.

6 To make a stock bottle, first gently shake the Mother Essence (follow your intuition) to awaken her healing gifts. Pause and give gratitude for what she offers. Place seven drops of the Mother Essence into a ½ or 1 oz. (15 or 30 ml) amber dropper bottle that contains 60 percent spring water and 40 percent brandy. Label and store in your medicine cupboard. Stock bottles are used to make the personal remedy bottle, which is the dilution taken by a person or animal.

7 To make a personal remedy bottle, shake the stock bottle to awaken the essence's healing gifts. Pause and give gratitude. Place two to four drops of one or more stock flower essences into a sterilized 1 oz. (30 ml) dropper bottle that is nearly filled with spring water. Some people add a few teaspoons of brandy, glycerin, or apple cider vinegar as a preservative; this is optional and depends on the season and the heat of your local climate. Place the dropper back into the bottle and label. Tap the bottom of the bottle with your fingers several times to awaken the life force of the flower essence. I like to place my hands in a prayer gesture, hold the bottle in my hands, and ask that this essence help in the healing process of myself or another being.

LEFT
Marigolds are a lovely addition to healing baths and footbaths.

RIGHT
Sunflower essence helps us to stand tall and radiant like the sun (refer to the beautiful Soulflower Plant Spirit Oracle Deck).

OVERLEAF
Separating rose petals from the calyx and placing in harvesting baskets

8 Another option is to place two or four drops from a stock flower essence bottle directly into a bath, oil, salve, or herbal tincture bottle. A flower essence can be taken for as many days or weeks as you feel drawn to take it. When I find myself taking a specific essence less often, this usually means I no longer need it. If there is still water in my personal remedy bottle and I feel completed with this essence, I take the bottle outside and offer the remainder back to the earth. I reuse my bottles and droppers after boiling them in clean water for five minutes.

Flower essences contain only the energy of the flower and no chemical constituents. This makes them safe during pregnancy and nursing and for infants, children, animals, and adults of all ages. If a strong emotional reaction occurs, seek emotional support from a friend or health care provider. Deep healing cannot happen in isolation or when shame or blame is present. Forgiveness, self-acceptance, and love help the soul and spirit heal.

Let's have tea. Let's have
galaxies, let's have earthworms,
let's have sorrow and tenderness,
and let us pour and receive the
bottomless mercy that life has for
us in our foolishness, our failures,
and our most secret longings.
In return, let us forgive the world for
being the world, let us allow all
things to be forgiven, to be blessed,
just for a moment, just for the
duration of a cup of tea.

—Joan Sutherland

Part IV

HEALING WITH HERBS

PREVIOUS
Preparing to harvest
a basket full of fresh
lavender flowers

ABOVE LEFT
Harvesting anise
hyssop

LEFT
A screen full of anise-
hyssop flowers and
leaves ready for the
drying room

ABOVE
Checking in on anise
hyssop seedlings

My mother's roots are in the Mediterranean region and northern Europe, hence my love for Mediterranean herbs. (My father's roots I do not know, as I was adopted.) This book is written from my personal experiences. I cannot speak for the bold and courageous women healers and herbalists whose rich and varied traditions, remedies, and rituals are different from mine. It is their place to do so.

The herbs I chose to include in this section are ones that are fairly easy to grow and that I have been growing, gathering, and preparing medicine from for several years.

Anise hyssop
Agastache foeniculum
Lamiaceae family

Place of origin: prairies and southwestern region of North America

Healing qualities: These beautiful purple flowers add color to wintertime tea blends. Their mild and pleasant licorice flavor relaxes the nervous system and eases indigestion. The fresh flowers and leaves create a lovely sun tea in summer. The glycerite is a yummy remedy for children and adults with tummy aches or indigestion.

Growing and gathering: Perennial. USDA zones 3 to 9. Anise hyssop's violet-blue flowers are favored by bees. Seed is light-dependent and germinates within two weeks if seeded indoors in a warm location. Gently tamp seed into organic potting soil and mist

Anise hyssop flowers

to keep moist. Prefers full sun and well-drained soil. Space plants 12 in. (30 cm) apart. They will reseed once established. If garden space allows, consider growing several plants as a hedgerow (pollinators appreciate hedgerows). This plant reaches a height of 36–48 in. (91–122 cm), with several branching, flowering stalks. Felco 310 clippers are my favorite tool for collecting flowers and leaves, and they work well for cutting the flowering stalks of anise hyssop. Gather the flowers when they first begin opening.

Drying: Lay stalks on drying screens, or bundle and hang. Once the leaves and flowers are dried (three or four days), strip them off the stalks and store in an airtight container.

Flavor: slightly bitter, sweet, pungent

Temperament: cooling

Preparations: fresh or dried leaf and flower tea; fresh glycerite; fresh vinegar; oxymel; freshly dried flowers in honey; cordial

Bergamot, wild
(also called sweet leaf, purple bee balm)
Monarda fistulosa
Lamiaceae family

Place of origin: North America

Healing qualities: Anishinaabe herbalist Keewaydinoquay taught her university classes that wild bergamot's main gift is to help colicky babies. I also appreciate this herb for relaxing nerves, soothing burns, treating winter colds and fevers, and helping to resolve acute and chronic urinary tract infections, systemic yeast, and herpes (with calendula, echinacea, lemon balm, and licorice root). Wild bergamot and the red flowering species (*Monarda didyma*), combined with calendula blossoms, add color, flavor, and healing benefits to wintertime teas and oxymels. The long, tubular, radiant red flowers make a special flower essence for encouraging us to fully step into our lives, celebrate each day, live our unique gifts fully, open our hearts, and sing or speak with ease and joy.

Growing and gathering: Perennial. USDA zones 3 to 8. Wild bergamot commonly reseeds in gardens, is easy to establish from transplants, and prefers sun and moist garden soil. Consider growing both *Monarda fistulosa* and *Monarda didyma* for pollinators and to make wintertime teas. The red flowering species, *Monarda didyma*, grows in partial shade and full sun. The flowers of *Monarda fistulosa* and *Monarda didyma* are important for bees, hummingbirds, and butterflies.

LEFT
Purple bee balm
growing in front of a
statue of Kuan Yin,
the Chinese name for
the Bodhisattva of
compassion throughout
the Buddhist world

ABOVE
Purple bee balm
and yarrow, a lovely
combination for a
healing bath

OPPOSITE
Red bee balm

The lavender-colored *Monarda fistulosa* flowers attract various native bees and butterflies. The red flowers of *Monarda didyma* are especially loved by hummingbirds; the species is known as Oswego tea. Oswego acknowledges the Native people of New York State, who enjoy this herb as a beverage tea. I keep a garden bench close to red flowering bee balm plants so I can sit and closely watch hummingbirds dart and feed on their beautiful flowers.

Drying: Lay on screens, or hang in loose bunches. Once dried (after two to four days), clip the flowers, strip the leaves off the stalk, and place in an airtight container.

Flavor: pungent, sweet, sour

Temperament: warm to hot, diffusive

Preparations: fresh or dried flower and leaf tea; fresh tincture; oxymel; leaf poultice for burns; flower essence; freshly dried flowers in honey

LEFT
Black cohosh flowering

ABOVE
Black cohosh in early
stages of growth in May

Black cohosh

Cimicifuga racemosa syn. *Actaea racemosa*
Ranunculaceae family

Place of origin: eastern North America

Healing qualities: Indigenous women of North America have long
relied upon black cohosh for female-related conditions and shared their
knowledge with European settlers. Currently, black cohosh is called

upon to reduce premenstrual stress with emotional moodiness and depression, painful menstruation, ovarian pain, labor and postpartum pain, and menopausal symptoms such as mild depression, moodiness, and hot flashes. The root's antispasmodic properties ease bronchitis, asthma, muscle spasms, and lower back pain. Please consider planting and sharing this important medicinal to honor Native American women for their generosity and to ensure this plant's survival.

Growing and gathering: Perennial. USDA zones 3 to 9. On our farm, black cohosh plants grow in several locations, some in partial to full shade and some in full sun. In regions warmer than Maine, they need full shade. As the climate warms, even in northern New England this medicinal may need full shade. The tall, white flowering stalks are magnificent and covered with native bumblebees when in full flower.

An important native woodland medicinal, black cohosh is listed on United Plant Savers' At-Risk list due to overharvesting and loss of habitat. It is easy to grow black cohosh from root divisions. Roots are best planted during early spring or fall, in well-drained soil rich in organic matter. Be sure the tops of the rhizomes are at least 2 in. (5 cm) below the soil's surface. In fall, we mulch our black cohosh beds with several inches of leaves.

Plants can also be started from seeds that have undergone a natural stratification process. (They need a month of warmth, then several months of cold in order to germinate.) Thickly sow freshly harvested seeds into a prepared garden bed in fall (mark well and mulch with leaves) or sow directly into wooden flats and place in a warm room for a month (keep soil damp) before placing the flat outside in a protected place for the winter.

We dig black cohosh roots in fall, waiting until they are three to four years old. The rhizomes grow in dense clumps, which I prefer to dig with a small garden fork, being careful to lift out the whole clump without breaking the rhizomes. I vigorously shake soil off these gnarly roots before taking them to be washed.

Drying: We hand wash our roots in a designated wheelbarrow with warm water, cutting them into larger pieces with heavy-duty Felco pruners (#8 or #9) to help remove any tightly compacted soil. Once they have been washed, we tincture the roots fresh or chop them into ½ in. (1.3 cm) pieces and lay on screens. The roots take six to seven days to dry.

Flavor: bitter, acrid; fresh root is sweeter

Temperament: cool

Preparations: dried root tea; fresh root tincture; flower essence. My friend Kate Gilday (woodlandessence.com) offers black cohosh flower essence for "knowing and trusting in one's inner strength and resources. For the honesty and courage to deal with and heal past experiences of abuse and oppression. Releasing entanglements. Bringing strong sense of self-emerging."

Safety considerations: Avoid use during pregnancy.

Harvesting black cohosh
root in the fall

Calendula

Calendula officinalis
Asteraceae family

Place of origin: southern Europe

Healing qualities: This magnificent medicinal flower warms and
uplifts the spirit, embodies the sun's light-filled nature, restores vitality, and heals damp and hidden places in the body. A valuable fall
and wintertime remedy, calendula adds joy and color to tea blends.
Calendula mildly stimulates the immune system and is beneficial for
reducing colds, swollen glands, flu, and fever (when combined with
wild bergamot and yarrow). The flower's bitter flavor gently stimulates
the actions of the liver and gallbladder, aiding the body in eliminating
toxins. Calendula also supports lymphatic detoxification following an
illness. A tea or diluted tincture soothes the mucosa (lining) of a person's mouth if it is raw or inflamed, and heals gum tissue after a tooth
extraction.

Calendula's antiseptic, antifungal, and anti-inflammatory actions
help resolve herpes, thrush, and vaginal infections and soothe irritated vaginal tissue. Calendula oil eases tender, swollen breasts and
improves overall lymphatic circulation. The oil and salve are safe and
effective for healing sore and cracked nipples for women who are
pregnant or nursing, as well as for applying to the perineum before and
after birth and for soothing and healing various inflamed skin conditions (cuts, chapped lips, dry skin, diaper rash, eczema, hemorrhoids).
For anyone who has experienced physical or sexual trauma, calendula

TOP LEFT
I use my thumb and forefinger to clip calendula flowers from the stem during harvest.

TOP RIGHT
Calendula flowers

RIGHT
Calendula and yarrow growing together

LEFT
Harvesting baskets

BELOW
Freshly harvested
calendula flowers

is a valuable healing ally when taken orally (as a tea, tincture, vinegar, or oxymel), applied topically (as an oil or salve), or placed in baths. Calendula promotes tissue repair: when used topically it speeds the healing of surgical wounds, softens and lessens scar tissue, and helps slow-to-heal wounds.

The energy behind our thoughts and words has power; calendula's warm, sunlike radiance helps us cultivate compassion and understanding when communicating. Call on calendula to help you speak what is in your heart in a balanced and considerate way, while at the same time listening deeply to what is being said by another.

Growing and gathering: This brightly colored flower grows easily as an annual. When purchasing seed, be sure it is the specific type for medicinal purposes. Calendula grows best in full sun and in well-drained soil with compost and straw mulch. If you direct-seed, weed and compost the bed ahead of time so the tiny seedlings can thrive without weed competition. Space seedlings 12 in. (30 cm) apart to give them plenty of room to bush out. Each plant will produce an abundance of new flowers every week when given space and healthy soil. Calendula will reseed in the garden, though the second generation may be slightly different from what you originally planted. When relying on plants that have reseeded, thin or move them around in your garden so they have ample space to grow. Replant them with compost and then place mulch around each plant. Calendula can also be grown in pots on a sunny deck.

On our farm, we sow calendula seeds into plug trays inside our heated greenhouse in late April. Seeds are lightly covered with ½ in.

(1.3 cm) of soil and germinate within four to five days. We bring these seedlings outside in late May to harden off (adjust to the outside cooler temperatures) for two to five days before transplanting them into beds that are prepared with compost and mulch. Mulching the bed with straw before transplanting seedlings saves time and retains the soil's moisture for the plants.

Calendula flowers need to be collected every two to three days, ideally on sunny days, once the dew is dry. This continuous picking keeps the plants producing flowers until the fall frost. I prefer to gather the flowers with my fingers, which is more efficient than using clippers. I also like feeling the flowers' sticky, antiseptic resin on my fingers as I gently snap the leaf stem at the base of each flower and carefully place the flower into a basket.

Drying: Calendula flowers dry best when laid on their side on a screen rather than upside down or upright. They take longer to dry than most flowers, at least seven to nine days. On our farm, we have found that 8 lb. (3.6 kg) of fresh calendula blossoms dry to about 1 lb. (0.5 kg). Give the process an extra day or two if in doubt. I run my hand through the dried flowers and squeeze numerous flower centers to make sure they are not damp or flexible before placing them in brown paper bags.

Flavor: bitter, slightly pungent

Temperament: warm, dry

Preparations: fresh and dried flower tea; fresh flower tincture, vinegar, oxymel, and sun tea; dried flowers in oils, salves, and creams; fresh

Calendula flowers laid out to dry on a screen in the drying room

or dried flowers in footbaths and healing baths; fresh or dried petals in honey or salads; flower essence

Safety considerations: Avoid oral use during pregnancy; topical use is safe. People with allergies to plants in the Asteraceae family may need to avoid both oral and topical use of calendula.

Grow a Row of Calendula: The Grow a Row project (helpgrowarow.org) is a grassroots initiative to encourage herbalists and gardeners to grow a row of calendula, make oil and salves, and donate them to organizations serving women and children. (See page 119.)

Dandelion

Taraxacum officinale
Asteraceae family

Place of origin: Asia and Europe; today, dandelion is commonly found worldwide in temperate regions.

Healing qualities: Dandelion roots and leaves are my favorite spring tonic food and medicine. The root's bitter flavor stimulates saliva and hydrochloric acid production, bile flow, and digestive enzymes, all of which improve sluggish digestion and the digestion of fats; ease indigestion, gas, and constipation; decongest the liver; and improve the body's detoxification pathways. As an overall liver tonic, the root and leaf can be taken daily over two to three months following a long-term illness or surgery or as part of a protocol for ameliorating the drastic hormone shifts and imbalances associated with premenstrual and menopausal stress. Dandelion leaf contains potassium and is an effective diuretic that is nonirritating to the kidneys and urinary tract. It relieves the fluid retention common before menstruation. Dandelion's cooling temperament smooths out the feelings of agitation and anger that can occur prior to menstruation, during menopause, or in other times of stress.

My friend Lisa Estabrook, painter and creator of the Soulflower Plant Spirit Oracle Deck, writes: "Let it go! Dandelion wants you to listen to the messages your body is sending you and release the attachments that no longer serve your soul's journey. Whatever you are hanging on to, she will help you detoxify one thought at a time,

releasing negativity and self-judgment, and restoring ease and flow to your life." Whether taken as a flower essence, eaten as fresh greens, drunk as a root tea, or used a whole plant tincture, dandelion is an herb to grow and to know. Dandelion is also an herb that begs us to stop the annual spraying of more than ninety million pounds of herbicides and pesticides on lawns in the United States (see my book *How to Move Like a Gardener*).

Growing and gathering: Perennial. USDA zones 3 to 10. Dandelions are easily found in gardens, along stone pathways, and at the edges and walkways of farmers' fields. They grow exceptionally well in full sun and in soil that is fertile, moist, and deep. In ideal conditions, the roots can be as long as 12 in. (30 cm).

Dandelion seed can be sown indoors and transplanted outside when three to four weeks old or seeded directly into well-prepared garden beds in either late fall or early spring. Space each plant 7–8 in. (18–20 cm) apart. In a bed 60 in. (150 cm) wide, you can plant three rows across, 15 in. (38 cm) apart. The whole plant can be dug in the fall of the first year, in the spring (before flowering), or in the fall of the second year. Let some plants flower and seed each year. The flowers' nectar feeds honeybees and migrating ruby-throated hummingbirds headed north in spring.

Dandelion greens are one of the most nourishing spring foods to enjoy in salads, soups, and vegetable dishes, or with eggs. They are most tender in spring; summer and fall greens are a bit tougher and more bitter but still valuable for putting in vinegars and oxymels. I use Felco 310 clippers when harvesting spring dandelion greens for food, taking a few leaves from different plants so each will continue to grow. I use a small garden fork when harvesting the whole plant for medicine, loosening the surrounding soil. This ensures that the root will lift out of the earth without breaking.

As with all roots, we shake them well while in the garden, returning as much soil as possible to the bed. We initially rinse the whole plants in buckets with lightly warm water and then place them in our yellow wheelbarrow for further washing. They are next placed on a screen and hosed again with clean water. Leaves are clipped off and

Freshly harvested dandelion root

washed separately in a bucket. We often let roots and leaves drip-dry on a screen for thirty minutes before bringing them inside to tincture.

Drying: Both leaves and roots can be dried. I slice the roots into ½ in. (1.3 cm) pieces with a sharp knife or Felco #8 or #9 pruners and lay them on a screen. Chopped roots dry within four to five days if the temperature is around 90–100°F (32–38°C). Approximately 4 lb. (1.8 kg) of root dries to 1 lb. (0.5 kg). Chopping roots before spreading them on a screen speeds the drying. Home gardeners who wish to dry the leaves can let them drip-dry outside in the shade for several hours, then use a clean towel to absorb excess moisture before laying them on screens to dry. Check leaves daily and bag them up before they are overly dry and crumble in your hand.

Flavor: bitter, slightly sweet

Temperament: cool, moistening

Preparations: fresh or dried roots in tea, tincture, or glycerite; fresh leaves in tincture, salads, soups, and vegetable dishes; vinegar and oxymel (fresh or dried root, fresh leaf and flower); flower essence. Flowers can be fried in a batter of flour, eggs, and milk or made into wine. Dandelion greens are a wonderful addition to pesto.

Safety considerations: Use cautiously if you are taking antihypertensive or diuretic drugs. Theoretically, a person allergic to plants in the Asteraceae family may have a reaction to dandelion flowers.

A jar of dried dandelion root

Echinacea

Echinacea purpurea, E. angustifolia, E. pallida
Asteraceae family

Place of origin: prairies of eastern and central North America

Healing qualities: *Echinacea angustifolia* remains to this day an important medicinal plant for the Great Plains Indians. They generously shared with European settlers their knowledge regarding the herb's efficacy in treating insect bites, poisonous snakebites, sore throats, abscesses, and septicemia.

For more than thirty years I have cultivated, harvested, and saved seed from the *E. purpurea* species and prepared hundreds of gallons of fresh root, leaf, and flower tincture. When I spend time looking closely at the spiral pattern on top of each cone flower, I am reminded to open my heart and be present with life's spiraling rhythm. I reach for echinacea tincture at the onset of a viral or bacterial infection, especially when the throat is sore, the body has been exposed to wind, and/or a thick yellow-greenish sputum (a sign of heat) is being coughed up. I learned from acupuncturist and herbalist Thomas Avery Garran that echinacea's acrid nature thins phlegm and sputum so they can be more easily expectorated from the body. The plant's bitter and cool nature clears the heat caused by the infection. Echinacea and burdock root, taken internally three to four times daily, is one of my favorite combinations for resolving boils, carbuncles, and lymphatic swellings. Diluted echinacea root tincture and calendula flower tea or tincture soothe canker sores, tonsillitis, and inflamed gums. Gardeners should keep a bottle of

A bee visiting
echinacea flowers

echinacea tincture close at hand during tick season. Put it directly on a tick bite and cover with a Band-Aid; repeat twice daily for two days.

Growing and gathering: Perennial. USDA zones 3 to 8. There are nine different *Echinacea* species. *Echinacea purpurea* is the easiest to cultivate from seed, either by direct sowing into prepared beds in fall or planting in plug trays in a warm, sunny window or greenhouse in spring. Once the seedlings are eight weeks old, we harden them off outside and transplant them into a sunny location with well-drained soil that has received compost and is fully mulched with straw. If you have enough room in your garden, consider grow-ing several echinacea plants (fifteen or more) in a bed together, as you will attract various pollinators and benefit from standing among the fragrant flowers, enjoying their beauty. Space seedlings 15–18 in. (38–46 cm) apart.

During the plants' second and third year, we walk among the rows of echinacea, gathering a few leaves and flowers from each plant using Felco 310 clippers. We grind the leaves fresh with alcohol and water, and we place the flowers whole into glass jars, then cover them with alcohol and water. In the fall, we dig our three- to four-year-old echinacea roots and tincture them fresh. Each of these tinctures macerates for six to eight weeks and is then poured off into individual jars. In early winter, we blend echinacea leaf, flower, and root tinctures together.

Drying: Washing *E. purpurea* roots requires extra effort because so much soil is embedded in the root mass. Quartering the roots before washing helps. Once the roots are washed, we chop or clip them into

Echinacea flowers, with mullein plants bordering

LEFT
Recently planted
echinacea seedlings

BELOW LEFT
Echinacea seeds
fourteen days into
the drying process

BELOW RIGHT
Echinacea flowers
have a special, sweet
fragrance that is
medicine for the soul.

OPPOSITE
Freshly harvested
Echinacea purpurea root

½ in. (1.3 cm) pieces and lay them on screens to dry, where 3 lb. (1.4 kg) of root will dry to about 1 lb. (0.5 kg) over the course of four to six days. Gather echinacea leaves in summer and lay them on screens to dry for winter teas.

Flavor: pungent, bitter, acrid

Temperament: cool, neutral

Preparations: fresh or dried root tea or tincture; fresh leaf and flower tincture; dried leaf tea; fresh or dried root glycerite; mature seed tincture; dried root in oil and salve; flower essence. Good-quality root tincture creates a tingling sensation in the mouth.

Safety considerations: Avoid taking echinacea flower tincture if you have allergic reactions to pollen from plants in the Asteraceae family (chamomile, daisy, ragweed). People with autoimmune conditions should seek the advice of a clinical herbalist before taking echinacea.

Hawthorn

Crataegus laevigata, C. monogyna, C. oxyacanthoides, C. phaenopryum
Rosaceae family

Place of origin: many different species native to Asia, North America, and Europe

Healing qualities: Hawthorn's energy offers us the courage to stand upright and live with a warm, steady, and open heart; to forgive, accept, and love ourselves and others; and to care thoughtfully for our energetic boundaries, while also understanding that we are part of this beautiful, shimmering web of life. Hawthorn soothes and strengthens the heart in times of grief and loss. When the emotional heart and spirit are supported, more peaceful sleep is possible. Hawthorn flower tea, tincture, essence, and syrup calm and settle the heart and spirit, lessen anxiety and agitation, and comfort children and adults who easily feel vulnerable or are tenderhearted.

Hawthorn, my favorite cardiovascular tonic, improves heart muscle function, lowers mildly elevated blood pressure, reduces inflamed connective tissues, and supports overall circulation. Hawthorn and lemon balm is a combination I commonly recommend for children and adults who live with attention deficit disorder and for their family members. Hawthorn, in combination with rose, lemon balm, lavender, and milky oats, supports the heart and nervous system of a person who has experienced trauma or loss.

Harvesting hawthorn flowers and leaves near Avena's majestic willow

Growing and gathering: Long-lived tree. Hardy to USDA zones 3 to 8. Hawthorn can be grown from seed, but growing from cuttings is much quicker. In 1997, I purchased fifty young Washington hawthorn trees, 24 in. (60 cm) tall, from the Fedco tree catalog and planted them 120 in. (300 cm) apart, creating a natural living fence around Avena's Sanctuary garden. Hawthorn prefers well-drained soil and will grow in full sun or at the edge of a forest in partial shade. We mulch our hawthorn hedgerow with wood chips and have encouraged the flowering perennial sweet cicely (*Myrrhis odorata*) to grow and spread under the trees, the large, lacy flowers and leaves providing nectar for honeybees and a living mulch.

Hawthorn hedgerows create magical spaces for birds, honeybees, and nature spirits. The protective thorns are 1–2 in. (2.5–5 cm) long and require us to be fully present when we engage with this tree. The nectar from hawthorn flowers is sought after by honeybees and other tiny native bees. We gather the leaves and flowers when the buds first open in mid- to late June. Using Felco 310 clippers, we collect what we can from the ground level and then climb a tripod-type wooden apple ladder to reach into the trees. The delicate flowers and leaves are tinctured immediately after gathering. In mid-October, once the berries are fully red, we again use Felco 310 clippers to harvest clusters of berries that are reachable from the ground or from the ladder. We use a long pruning pole for cutting the higher branches that hang heavy with berries. Once the branches are on the ground, we use the Felco 310 to clip the clusters of berries into baskets. The fresh berries are tinctured soon after being harvested.

Drying: Leaves and flowers, laid on a screen, dry within two to three days. Once they are dry, I store them whole in airtight glass containers. The berries are fleshy and take much longer to dry. I lay a single layer on screens for a minimum of two weeks. Their dryness can be tested by nibbling on several berries and by cutting them with the Felco 310 clippers. When they are hard to cut and not flexible or moist in any way, they are dry. I keep the red berries in brown paper bags for a few weeks to ensure they are totally dry and then store them in glass jars for wintertime teas.

Flavor of flowers: sweet, slightly bitter

Flavor of berries: sour

Temperament: slightly warm

Preparations: fresh or dried flower and leaf tea; fresh or dried berry tea; fresh leaf and flower tincture; fresh or dried berry tincture, oxymel, or syrup; dried berry powder; flower essence

Safety considerations: Seek professional support if you are taking pharmaceutical cardiac medications and wish to use hawthorn.

TOP LEFT
Preparing for the autumn hawthorn berry harvest

BOTTOM LEFT
Freshly harvested hawthorn berries

TOP RIGHT
Hawthorn berries on a drying rack

BOTTOM RIGHT
Hawthorn tincture made from fresh berries ground in alcohol and spring water

LEFT
Holy basil

ABOVE
Harvesting holy basil

Holy basil / Sacred basil / Tulsi / Tulasi
Ocimum tenuiflorum, O. sanctum
Lamiaceae family (formerly Labiatae)

Place of origin: India, Afghanistan, Pakistan

Healing qualities: Many households in southern India tend a Tulsi plant in their courtyard for divine protection and clearing negativity. Several years ago, while a guest in a small village in India, I arose each dawn and walked through the streets, quietly observing villagers offering rice, water, incense, marigold flowers, and prayers to the Tulsi plant in their courtyard. This magnificent plant has been honored as an embodiment of the Divine Mother for a very long time in southern India. Her spiritual gifts include opening the heart and mind, inspiring harmony and spiritual clarity, enhancing feelings of love and compassion, and guiding humans to live consciously and with reverence for all of Nature.

Tulsi is highly valued in the Ayurvedic system of healing for preventing and reducing fever and flu symptoms, relieving sore throats, clearing congestion from the lungs and upper respiratory tract, and increasing lung capacity. Many herbalists call upon holy basil for supporting immunity and enhancing overall balance, clarity of mind, and well-being.

The aroma and spirit of this plant open the heart and strengthen our inner capacity for compassion and resilience. Meditating quietly near a holy basil plant in the early morning is a practice that helps me connect with my breath and heart and enhances my devotion to the healing plants and their pollinators.

Dried red clover, holy basil, and lady's mantle leaf and flower

Growing and gathering: Short-lived perennial in southern India and USDA zones 7 to 9. In cooler climates, like Maine, Tulsi grows as an annual and is easy to start indoors from seed. Lightly cover the seed with soil and keep moist in a consistently warm space or greenhouse for germinating. On our farm we transplant young seedlings into 3–4 in. (8–10 cm) pots when they are three weeks old. This helps them develop a stronger root system. We wait until early June, when the soil is warmer and there is no sign of frost, to plant seedlings in the garden.

CLOCKWISE FROM TOP LEFT Harvesting holy basil

Transplanting holy basil seedlings into bigger pots

Gardener Anna Leavitt transplanting hundreds of holy basil seedlings into larger pots in Avena's greenhouse

Holy basil needs moist, well-composted soil and full sun to thrive.
We cover the bed with straw before transplanting, and leave about
12 in. (30 cm) of space between seedlings. You will have happy bees
if you grow a bed of holy basil with twenty-five or more plants.
Tulsi also grows well in large pots placed outdoors on a sunny deck.

The plants get big and bushy and provide an abundance of leaves
and flowers for medicine when their tops are regularly cut; we clip
them once or twice a week throughout the summer using Felco 310
clippers. Ideally, prepare a fresh glycerite or tincture from holy basil
as quickly as possible after harvesting. Organic glycerin draws out the
valuable essential oils, creating an uplifting and delicious holy basil
glycerite.

Drying: Lay out the flowering tops carefully and singly on screens
to maintain their quality during the drying process. They usually take
seven to ten days to completely dry. Store as whole as possible in
airtight glass jars.

Flavor: bitter, pungent

Temperament: dry, warm

Preparations: fresh or dried leaf and flower tea; sun tea; fresh
tincture, glycerite, oxymel, or syrup; dried powder; honey; fresh or
dried in footbaths and healing baths; hydrosol; essential oil

Safety considerations: Avoid use during pregnancy.

Lavender

Lavandula angustifolia
Lamiaceae family

Place of origin: native to the mountains of the Mediterranean

Healing qualities: Lavender is a remarkable herb for easing tension and gas in the digestive system, stress, stress-induced headaches, anxiety, mental and emotional agitation, insomnia, and feelings of unrest and inner disharmony. Lavender's antiseptic and anti-inflammatory

LEFT
Harvesting lavender

RIGHT
Gardener Brittany Cooper with a basket full of freshly harvested lavender

LEFT
Giving gratitude to
the lavender before
harvesting, surrounded
by lavender with
milkweed, mullein,
and black cohosh

ABOVE
Native bumblebees and
honeybees all adore
lavender blossoms.
Consider growing a
hedgerow of lavender
for attracting bees, for
making medicine, and
for napping near.

properties soothe sore throats, inflamed gums and mouth sores, tonsillitis, and laryngitis, and help resolve bronchitis and pneumonia. Steam inhalations made with a few drops of pure lavender essential oil support the respiratory and nervous systems. A few drops of essential oil or a tea applied topically are effective for healing burns, infected cuts, insect bites, and red, irritated skin conditions. Lavender is an outstanding, accessible herb for calming the mind and uplifting the spirit.

Growing and gathering: Woody perennial. Hardy to USDA zones 4 to 9. Lavender is easiest grown from cuttings or seedlings, though with patience this plant can be grown from seed. Prefers sandy, well-drained soil and full sun. In Maine, we grow the hardy Munstead variety in hedgerows, spacing them 15–18 in. (38–46 cm) apart. Plants are pruned once they reach two to three years old. A fall pruning encourages healthy growth the following year and reduces plant stress during winter. We carefully leave 2–3 in. (5–8 cm) of silvery-green growth above the plants' woody stems when pruning. In the winter, we mulch our plants with a heavy, breathable material called Reemay.

The ideal time for gathering lavender is when the flowers first open. We gently hold several lavender stalks in one hand and cut them just above the leaves with the Felco 310 clippers, then carefully lay the bundles in baskets, facing the same direction. These bundles are immediately prepared into fresh lavender glycerites or laid out to dry.

Drying: When placing them on screens, lay the stalks close to one another in a single layer. Laying the flowers in the same direction makes it easier to put them away. The flowers will dry within a few days. I wait a few extra days until the stalks are dry, as I use both the

Freshly harvested
lavender laid out to dry

stalks and the flowers in my winter teas and footbaths. The flowers
easily rub off the stalks for storage in glass jars.

Flavor: bitter

Temperament: cool, dry

Preparations: fresh or dried blossom tea; fresh or dried blossom
tincture, glycerite, vinegar, oxymel, or infused oil; baths (foot, sitz,
healing); essential oil; hydrosol; culinary

Lavender or Lemon Balm Lemonade
Make an infusion from fresh lavender blossoms or lemon balm leaves.
Leave covered as the tea water cools. For every cup of tea, add ¼- to
½-cup (60–120 ml) raw honey and fresh organic lemon juice to taste.
Keep cold in a refrigerator no longer than three days. Enjoy drinking
this delicious herbal lemonade on a hot day.

Lemon balm

Melissa officinalis
Lamiaceae family

Place of origin: Mediterranean

Healing qualities: My favorite ways to enjoy fresh lemon balm are in tea or glycerite. *Melissa* eases anxiety, panic attacks, insomnia, attention deficit disorder, and palpitations associated with stress; reduces feelings of being overwhelmed; and supports people with an overactive thyroid. Lemon balm helps prevent the breakdown of acetylcholine, a neurotransmitter important for cognitive function. This gentle herb is commonly used for lowering fevers; preventing and treating outbreaks of chicken pox, shingles, and herpes; easing insomnia during menopause; calming a nervous stomach; and uplifting the spirit.

Growing and gathering: Herbaceous perennial. Hardy to USDA zones 4 to 9. Lemon balm is easy to start from seed: gently tamp the seeds into moist soil in a pot or plug tray, as they need light to germinate. Keep the seeds moist by misting. I transplant three or four tiny seedlings into a 3 in. (8 cm) pot and let them grow into a small clump before replanting in the garden. Lemon balm prefers a mostly sunny location, well-drained soil, and straw mulch. Space about 12 in. (30 cm) apart. It is lovely to grow a bed of several plants, as they will form a thick mat of fragrant leaves.

Felco 310 clippers work well for nipping a few leaves for daily tea. When harvesting dozens of pounds of lemon balm, we use Felco

clippers or a *kama* (a sickle-like tool) to cut the whole plant, leaving 5–6 in. (13–15 cm) above ground to support ongoing growth. The plants are immediately prepared into a fresh glycerite or tincture or laid out to dry.

Drying: The essential oils in the leaves are delicate. To preserve lemon balm's quality, dry at lower temperatures, around 70–75°F (21–24°C), with the leaves placed on racks close to the floor. Good air circulation with fans quickens the process. As soon as the leaves are dry (two to four days), strip them from their stalks and store in a glass jar.

Flavor: sour, mildly bitter

Temperament: cool, moist

Preparations: fresh and dried tea; sun tea; fresh tincture and glycerite; infused oil; salve; foot and healing baths; hydrosol; essential oil

LEFT
Deeply enjoying the fragrance of fresh lemon balm

RIGHT
Moving lemon balm in preparation for weighing and drying

Lemon verbena

Aloysia citrodora
Verbenaceae family (Verbena)

Place of origin: Argentina, Chile

Healing qualities: This fragrant, lemony herb is uplifting to the spirit, eases stress, and soothes digestion, gas, and colic. Lemon verbena cleanses negativity and restores energy when combined with other herbs (holy basil, lemon balm, lavender) into a foot or healing bath. Weekly aromatic baths are simple to create, and when combined with a healing prayer, they profoundly shift fears, tensions, and outdated beliefs.

Growing and gathering: Perennial, woody shrub in tropical regions, native to Peru and Chile. Hardy perennial in USDA zones 8 to 11. Easily propagated from cuttings. For more than twenty years, I have wintered several lemon verbena shrubs in plastic pots in a sunny room. We bring the plants outside in late April (weather permitting) and back inside in late October. I let them acclimate for a week outside in the spring before removing their pots and planting them in well-drained soil in full sun. Lemon verbena also does well in pots on a sunny deck.

The leaves grow in whorls on semi-woody stems. Throughout the summer, I clip 3–4 in. (8–10 cm) leafy stems for fresh tea and baths. Cut just above a set of whorled leaves so that new stems will emerge. When fall days are sunny and warm, before the frosty nights arrive, I cut my plants back by a third—some of the woody stems will be two

feet in length—to promote bushier plants the following summer.
The plants go dormant in winter. I carefully water them once or twice
a week, enough to keep the soil slightly damp but not too wet, as they
prefer dryer soil. If you are not going to winter your plants, cut the
whole plant to dry.

Drying: Lay woody stems with leaves on a screen. The leaves dry
quickly, within a few days. You can then strip off the dried leaves and
lay them on the screen for one more day to ensure they are totally
dry. Store in an airtight container. Dried lemon verbena's lemony flavor
lasts much longer than that of dried lemon balm.

Flavor: pungent, slightly bitter

Temperament: neutral to slightly warming

Preparations: fresh and dried tea, syrup, glycerite, oxymel; freshly
dried in honey, cakes, sauces, or chocolate; foot and healing bath;
hydrosol

Marshmallow

Althaea officinalis
Malvaceae family

Place of origin: Europe

Healing qualities: Mucilage in the roots is beneficial for soothing dry coughs and dry sore throats, breaking up congestion and hardened mucus in the respiratory tract, easing inflammation in the digestive and urinary tract, and supporting the healing process for gastric and duodenal ulcers. I like adding marshmallow root elixir into blends that ease chronic urinary tract infections and moisten dry tissues, especially for postmenopausal women. A tea of marshmallow and mullein leaves combined soothes respiratory conditions. Marshmallow's soft, velvety leaves and delicate flowers are healing remedies for people whose hearts are hardened from life's challenges or those whose grief has interfered with their ability to feel their emotions. I like to stand near a flowering marshmallow plant and gently rub the leaf on my face. Ruby-throated hummingbirds, much to my delight, often feed on the plant's flowers.

Growing and gathering: Herbaceous perennial. Hardy to USDA zones 3 to 9. This silvery-leafed perennial with delicate flowers is easy to grow from root divisions and seed, and reseeds abundantly in the garden. The plant will reach 48–75 in. (122–191 cm) in height. Keep an eye out for tiny, dark-purple leaves poking through the soil in early spring. When planting seeds indoors, gently rub the seed

BELOW
Marshmallow root

RIGHT
Marshmallow flower

Sign in Avena Botanicals driveway with dozens of yellow flowering Greek mullein plants

with medium-grit sandpaper (a process called scarification) before sowing in pots.

Seedlings from the greenhouse are transplanted outside when they are five to six weeks old and spaced 15–18 in. (38–46 cm) apart. Marshmallow prefers damp garden soil and partial-to-full sun. Snip a few leaves from each mature stalk with Felco 310 clippers for making tea. Roots are whitish and fibrous. They can grow to be 24 in. (60 cm) long and are harvested once the plants are three years old. We dig roots early in the spring, as the first leaves begin to grow, or in the fall, once the plants have gone to seed. A fork and spade work well for digging; take care to lift the long roots out as unbroken as possible. The crowns can be replanted back into the garden after cutting off the roots. Fresh roots are made into elixirs or cut into ½ in. (1.3 cm) pieces for drying.

Drying: Leaves laid on screens in a single layer dry within three or four days. Chopped roots laid on screens take five to seven days to dry.

Flavor: sweet, slightly bitter

Temperament: moist, cold

Preparations: fresh or dried leaf and root tea; fresh leaf and flower sun tea; fresh or dried root elixir or syrup; flower essence

Mullein, Greek

Verbascum olympicum
Scrophulariaceae family

Place of origin: Europe

Healing qualities: The multiple flowering stalks of Greek mullein add a magical yellow glow to the garden when in full bloom. Not only humans, but also hundreds of bees benefit from this nectar source. Years ago, I created a fresh mullein flower elixir for a person whose lungs were deeply embedded with mucus. After a few days of taking the elixir alongside mullein leaf tincture and tea, the mucus began to loosen and move out of the lungs. Mullein flower elixir soothes lung tissue and expels phlegm. Mullein leaves also soothe the mucosa (lining), open the lungs and sinuses, and reduce coughing and constriction. Mullein flower oil, combined with calendula and garlic oil, soothes inflamed and infected ears (avoid using oil if the eardrum is perforated). Standing near a tall mullein flower stalk, or even touching the stalk with one's back, gives energy to those who have experienced trauma (old or current, personal or collective) and are unable to stand tall and feel confident within themselves. Mullein's magnificent flower stalk radiates a gentle but strong healing light.

Growing and gathering: Biannual. Hardy to USDA zones 3 to 9. Mullein can be grown from seed, and reseeds prolifically in the garden. Small seedlings transplant easily and can be given as gifts. Mullein prefers full sun and well-drained soil. The first-year plant grows into

**CLOCKWISE
FROM TOP LEFT**
Preparing to place an enormous dried mullein stalk in an area where a large compost pile will be built. We use dried stalks of plants as the base of our compost to help aerate the pile.

Walking through a stand of mullein in autumn

Mullein flower petals that have fallen on the mullein plant's leaves

Mullein flowers with a native bee

a lovely rosette of leaves, which can be harvested during the summer months for fresh tinctures or teas using the Felco 310 clippers. The root is dug in the fall of the first year and made fresh into a tincture. *Verbascum olympicum* roots are big; loosening them from the soil requires a long and extra-sturdy spade. The delicate yellow flowers, beloved by bees, are best gathered in the early morning (once dew has dried) and placed immediately in olive oil, organic glycerin, or honey. Early morning is my favorite time in the garden for observing the birds, pollinators, and plants waking up and then quietly gathering soft mullein flowers. New flowers emerge daily, and the old ones gently drop to the stalks' lower leaves.

Drying: Lay the leaves out in a single layer on screens for drying. The leaves of *Verbascum thapsus* dry faster than *Verbascum olympicum* leaves, whose midrib is thicker and requires more drying time (four to six days). I prefer to grow *Verbascum olympicum* because the multiple-flower stalks provide loads more nectar to the bees than the single-flowering stalk of *Verbascum thapsus*.

Flavor: salty (leaf), sweet (flower), bitter and sweet (root)

Temperament: moist, cool

Preparations: fresh or dried leaf tea; fresh or dried leaf tincture; fresh root tincture; fresh flower tea, glycerite, or elixir; fresh flowers in honey; fresh flower oil; flower essence

Nettle
Urtica dioica
Urticaceae family

LEFT
Harvesting nettle leaf with gardeners Anna Leavitt and Brittany Cooper

RIGHT
Nettle harvest on herb-gathering cloth with *kama* and protective glove nearby. After harvesting, gardeners gather up the cloth into a pouch and carry the nettle inside to be weighed and either dried or immediately made into fresh tincture or vinegar.

Place of origin: Europe; naturalized in temperate climates worldwide

Healing qualities: Nettle is one of the most nourishing herb and food sources on our planet. The leaf is high in minerals and micronutrients, especially iron, calcium, magnesium, potassium, and protein. Nettle leaf and seed teas restore the body's overall energy, offer nourishment to the adrenal glands, and (alongside lifestyle adjustments) support people who feel chronically tired. Nettle seed helps improve low thyroid function and strengthen the kidneys. The leaf supports the body's elimination of metabolic wastes and uric acid, and can be helpful in preventing urinary stones. Regular use of the leaf tea, fresh leaf tincture, or freeze-dried capsules relieves seasonal allergies, especially when this practice is started a month before allergy season begins. A quart of fresh nettle leaf tea, taken daily in spring for about a month, is my favorite spring tonic for renewing energy.

Nettle is a powerful teacher and healer for people who have experienced trauma and for folks who are highly sensitive. I fully agree with the English herbalist Nathaniel Hughes's insightful comments:

> Early trauma can lead to a hypersensitivity, as if one is always on the "look-out," a manifestation of deep rooted fear. This hypersensitivity, if used with skill, can be a great aid to empathy and healing, but without training can create confusion about what is self and what is other. This has the consequence of both compromising the sense of self whilst unconsciously internalizing the emotions of those around us. For such people, healing lies in the conscious creation of self-identity and the firming of boundaries. . . . One pattern I see in people with chronic illness is a struggle with the sense of their own edges, the boundary of self. They are constantly reacting to their environment, leaving insufficient energy to maintain their own integrity.

Nettle's horizontal rhizomes create a thick mat a few inches under the soil line. I imagine these rhizomes energetically spreading inside the body, restoring health to our tissues and cells, resilience to our soul, and loving boundaries to our whole body.

Growing and gathering: Herbaceous perennial. Hardy to USDA zones 2 to 10. Nettle thrives in moist, fertile soil high in organic matter and in old compost piles or barnyards. Nettle also improves soil fertility, adding nutrients to the soil and improving soil texture. Prefers

Nettle on a drying rack

partial to full sun. On our farm, we place tree bark chips between the rows of nettles to maintain walking paths. Nettle can be started from seed, though I have found that establishing a nettle bed from root cuttings to be the easiest propagation method. Either place the roots in pots with organic soil and keep them watered for several weeks before planting in the garden, or, if directly moving roots from one bed to another, ensure the new bed has plenty of compost and mulch. Space plants about 12 in. (30 cm) apart and water regularly while they are rooting themselves into their new home.

Most herbalists who live in the northeast eagerly await the first signs of nettle leaves emerging in spring for making tea and soups. Felco 310 clippers work well for nipping leaves into a basket for fresh tea. We do our big harvest of spring nettles, for tincturing and drying, when the plants are about 12 in. (30 cm) high and before they flower. To prevent being stung by the hairs on the stem and leaves, we wear long-sleeved shirts and rose-pruning gloves and use a *kama*. Collect the seeds when the green flowers begin to turn brown and form seed. Clip the whole flower stalk into a large bowl. These seeds can be tinctured fresh.

Drying: Place nettles in single layers on screens. The leaves dry within a week and can be stripped off the stems using gloves and a mask to prevent their tiny hairs from irritating the lungs. Alternately, rub them on a garbling table (a wooden frame or table made with ¼- or ½-inch stainless wire mesh) to separate the leaves from the stems. Store in airtight containers. To dry the seeds, lay them on top of a thin cotton cloth placed on a screen.

Flavor: salty, slightly sweet

Temperament: dry, cool

Preparations: fresh or dried leaf tea; seed tea; fresh leaf, root, or seed tincture; fresh leaf juice; fresh or dried leaf vinegar; fresh leaf infused overnight in olive oil; dried powdered leaf; fresh leaf lightly cooked as green vegetable; crushed seed in food. Fresh or dried root tea simmered in almond milk is an overall restorative tonic.

————

Oat seed, milky

Avena sativa

Poaceae family

Place of origin: Fertile Crescent in the Middle East; grows worldwide in fields and farmlands in both temperate and cooler temperature zones

Healing qualities: A fresh, green milky oat seed glycerite, tincture, or tea restores strength and vitality to the nervous system. This valuable nervous-system tonic, when taken over several weeks or months, eases mild to extreme anxiety, nervous debility, fear, fatigue, depression, mental restlessness, heart palpitations, premenstrual and menopausal stress, and (alongside lifestyle shifts) physical and mental exhaustion. Oats contain constituents that restore health to the myelin sheath (which covers the axons of the nervous system) and support people who feel emotionally drained and easily reactive. Oat seed glycerite helps reduce the cravings and agitation people experience when withdrawing from nicotine, caffeine, alcohol, drugs, or addictive behaviors. Oat is soothing and nourishing for people living with chronic degenerative conditions, and calming and rejuvenating to family members and caregivers. Oat's sweet gift helps relax tension and calm the spirit.

Growing and gathering: Oat is an annual grass. Hardy to USDA zones 2 to 10. Prefers well-drained and fertile soil and full sun. Scatter seeds across a bed prepared with plenty of compost and gently rake the seeds into the soil. A thick green carpet of oat sprouts will appear in

CLOCKWISE FROM TOP LEFT
Squeezing the milk out of milky oat seed

Milky oat seed

Harvesting milky oat seed

seven to ten days, especially if the bed is kept moist. Lightly cover the bed with straw or protective Reemay (a breathable cloth) after planting seeds to deter chipmunks, birds, and wild turkeys. Oat is a wonderful cover crop to plant in a different garden bed each year, giving the garden soil an opportunity to rest and be renewed. Once the seeds have been harvested as described below, the plants can be cut down with a hand sickle or scythe and gently forked into the soil. As they decompose, they add organic matter to the soil.

Watching the oat plant grow and the seeds swell and turn a greenish-yellow color is a practice in patience and faith. I begin checking the oat seeds for their milk content about two months after planting. The particular sound of the wind rattling the ripened seeds quiets my nervous system. As soon as the seeds are green and plump and a white, milky substance bursts from several squeezed seeds along the whole stalk (not just the ones at the top), I begin the harvest. I strip the seeds with my hands and place them into baskets, being careful not to overfill each basket, which would cause the seeds to sweat and their quality to diminish. Make fresh milky oat seed glycerite and tincture within a few hours of harvesting.

Drying: Place the seeds in a single layer on screens. We pick out the plant's green chaff by hand. The chaff can be dried separately and enjoyed as tea. Keep the room's temperature around 80°F (27°C); if it is too hot, the seeds' outer coat will dry too fast, before the inner part does, and when bagged up, spoilage will occur. In Avena's drying rooms it takes about eight to ten days, with good air circulation and consistent temperature, for the seeds to completely dry. About 4 lb. (1.8 kg) of fresh seeds dry to 1 lb. (0.5 kg). Check the seeds for moisture

Milky oat seed laid out
to dry

by cutting several of them from different drying racks with small
clippers. The inner white milk should crumble in your hand. Place
dried seeds into brown paper bags for a week before storing in glass
containers.

Flavor: sweet

Temperament: warm to neutral

Preparations: fresh or dried seed tea; fresh seed tincture, glycerite,
vinegar, or oxymel

Rose

Rosa rugosa, *R. damascena*, *R. centifolia*, *R. gallica*
Rosaceae family

Place of origin: *R. rugosa* originated in China

Healing qualities: Rose's delicate nature and uplifting fragrance offers joy, inner peace, love, and compassion to the heart. A special elder in my life, Ray Reitze, says, "Life moves through love." Taking a few drops of rose petal elixir, or smelling rose, helps open the heart and mind and inspires feelings of love and compassion for oneself and others. Roses are soothing and comforting in times of grief and loss. They alleviate stress, anxiety, fear, anger, and emotional agitation. Their antimicrobial properties reduce hyperacidity, heartburn, and diarrhea and clear toxins from the gut. Their cooling and decongesting qualities relieve uterine pain and spasms, reduce heavy menstrual and postpartum bleeding, and ease premenstrual and menopausal stress, including feelings of low self-esteem. Rose relaxes the nervous system, uplifts the spirit, and guides us to be loving and kind to ourselves and others.

Growing and gathering: Perennial shrub. Hardy to USDA zones 2 to 7. Roses prefer full sun and well-drained soil. They are easiest to grow from plants or transplanted runners early in the spring. Mulch with straw, organic buckwheat hulls, or seaweed (if you live near the ocean). We prune our *Rosa rugosa* hedgerow early in April by cutting the plants back by half. This encourages the roses to produce an abundance of flowers in late June.

Gathering roses early in the morning is a gentle and calming way to begin the day. Felco 310 clippers work well. If there have been several foggy or rainy days when we couldn't collect the flowers, I nip the spent blossoms to encourage the plants' further blooming. Roses bloom continuously for two to three weeks. The whole flowers or petals can be immediately prepared into glycerite, elixir, tincture, oxymel, or honey, or laid out to dry. In October, when the rose hips turn bright red, they can be collected for food, tincture, syrup, or drying.

Drying: Enjoy smelling the roses as you lay the whole flowers or petals on screens. Depending on the weather, they can take at least five to six days to dry. To dry rose hips, cut them in half and lay on screens or place in a food dehydrator at 100°F (38°C). They dry within a day in a dehydrator.

LEFT
Sitting with Avena's *Rosa rugosa* hedgerow, expressing gratitude to the plants before beginning the harvest

RIGHT
Greeting a *Rosa rugosa* flower before I begin to gather a basket of flowers

ABOVE
Harvesting *Rosa rugosa* from Avena's biodynamic hedgerow, with barn and drying room facility behind

LEFT
Separating rose petals from the calyx and placing them in distinct harvesting baskets

Flavor of flowers: bitter, pungent, slightly sweet

Temperament: cool, moist

Preparations: fresh or dried whole flower or petal tea; fresh petal or whole-flower tincture, glycerite, elixir, syrup, oxymel, honey, hydrosol, essential oil, footbath, healing bath, flower essence

OPPOSITE
Harvesting *Rosa rugosa*

CLOCKWISE
FROM TOP LEFT
Cutting rose hips in
half to prepare them
for drying

Harvesting rose hips
in the fall; a few rose
blossoms remain.

Rosa rugosa flower
petals and calyx

BELOW
Freshly harvested
Schisandra berries

RIGHT
Tending the
Schisandra vines

Schisandra
Schisandra chinensis
Magnoliaceae family

Place of origin: northern China, Korea, and Japan (twenty-four known species); *Schisandra coccinea* is a rare species found growing in undisturbed streambeds in North Carolina, Tennessee, Georgia, Florida, Arkansas, and Louisiana.

Healing qualities: Fresh *Schisandra* berry tincture is my favorite restorative tonic for building emotional and spiritual resilience. In summer, I place one dropper of tincture into my quart water bottle and sip throughout the morning. In fall and winter, I take a syrup made with hawthorn and *Schisandra* berries once or twice daily.

Chinese practitioners work with *Schisandra* when treating the lung, heart, and kidneys (this approach is different from Western medicine) and when helping people retain what is eternally valuable in their life. Western herbalists call upon this berry to strengthen the immune system and to ease chronic stress, anxiety, insomnia, poor memory, depression, fear, incontinence, and menopausal sweating. *Schisandra*'s astringent nature is supportive during major life transitions such as the death of a family member, animal companion, or close friend; physical or emotional transitions such as menopause; or the process of healing after trauma. Traditional Chinese herbalists say *Schisandra* stabilizes lung *qi* and is used for various respiratory conditions. The acupuncturist and Chinese herbalist Amy Jenner says:

The Heart in Chinese Medicine is "Lord and Sovereign" and is the residence of the Spirit (*Shen*, your Divine nature). If the Lord and Sovereign of your Body/Mind/Spirit is to bring your truest self into Life, it must provide an inner sanctuary from which the Spirit can shine its light. *Wu Wei Zi* (*Schisandra*) holds the Heart steady so your light can shine.

Growing and gathering: Perennial woody vine. Hardy to USDA zones 4 to 9. This sturdy vine prefers well-drained and deeply cultivated sandy soil, partial shade, plenty of moisture, rich compost, and tree-bark-chip mulch. The easiest way to cultivate is from root runners or layering. Three- or four-year-old plants will begin to send out multiple runners, especially if the original plants are mulched with tree-bark chips. Cut 6–8 in. (15–20 cm) of a root runner and plant in a large pot or prepared garden bed, then water regularly until the plant begins to sprout. When rooting into pots, keep them in a partially shaded area for one year. This helps them to establish a strong root system. When rooting in the garden, be sure to keep them watered daily and watch for their new green sprouts.

Schisandra vines need to grow on a strong trellis or arbor. The small, lightly fragrant white flowers grow in clusters, attracting thousands of small bees in early June. Female and male flowers grow on separate vines, planted 36 in. (91 cm) apart. Both give way to beautiful red bunches of berries in the fall. Vines begin fruiting after their third or fourth year. Berries are ready to harvest when deep red and juicy. We usually do two or three harvests, as not all the berry clusters ripen at the same time. Berries are tinctured fresh, and some are frozen for future fruit smoothies (combined with frozen organic blueberries, northern hardy kiwis, and bananas).

Schisandra tincture: fresh *Schisandra* berries ground in organic alcohol and spring water

Drying: Juicy fruit takes about two weeks to dry. Lay the berries out in a single layer on a screen. I check them for dryness by chewing them; fully dried berries do not open easily. Consistent, warm heat and air circulation from fans supports the process.

Flavor: sweet, sour, salty, pungent, bitter; the Chinese name, *Wu Wei Zi*, means "five flavors fruit"

Temperament: warm, dry

Preparations: fresh or dried berry tea; fresh berry tincture, syrup, or elixir; fresh juice; jam; frozen for smoothies; dried powdered berries; flower essence

Safety considerations: Avoid during acute fever, flu, bronchitis, pneumonia, or any acute conditions with excess heat such as heat rashes. Contraindicated for people and animals with epilepsy or anyone taking a phenobarbital or barbital medication.

Teasel

Dipsacus sylvestris
Caprifoliaceae family

Place of origin: Europe; naturalized in North and South America

Healing qualities: In 2001, I observed a mourning dove with a broken wing living in Avena's herb garden. Each morning, over several weeks, she emerged from a sheltered area and drank the water held within the leaves of the teasel plant. One day she flew upward to the nearby power line. She had healed! Later that summer, baby mourning doves appeared in our garden. This experience led me to study teasel in *The Book of Herbal Wisdom* by Matthew Wood. Since that time, I have included teasel root tincture in herbal blends for broken bones or inflamed and injured muscles, joints, and tendons, and as part of a protocol for people living with Lyme disease. I also recommend a few drops of root tincture or flower essence for people whose spirit feels broken—who feel emotionally exhausted, depleted, discouraged, vulnerable, or otherwise at odds with the world. According to David Dalton's *Stars of the Meadow*, "Teasel helps one find the way back to living in harmony with one's soul."

Growing and gathering: Biennial. Hardy to USDA zones 3 to 8. Scatter seeds in the fall in a prepared bed, or start indoors in pots in early spring. Transplant seedlings outdoors when they are six weeks old. Teasel prefers sun and well-drained soil. The roots grow bigger if the plant is not overcrowded; space 12 in. (30 cm) apart. Seedlings grow

Teasel cone heads. Their delicate purple flowers have passed, and now the seeds are developing within the brown cone heads.

well if watered during the first week after being transplanted. Dig the root in the autumn of the first year with a small garden fork or spade.

Carefully clip the prickly leaves from the white roots before washing them. In the summer of teasel's second year, a tall stalk emerges, with sharp leaves and several large, thistle-like flower heads. Tiny, lavender flowers encircle each flower head, attracting native bumblebees. Let a few plants flower each year and naturally scatter their ripe seeds, providing you with hundreds of tiny seedlings the following spring.

Drying: Chop into ½ in. (1.3 cm) pieces and lay them on screens or in a basket to dry for tea.

Flavor: bitter

Temperament: slightly moistening

Preparations: fresh root tincture; fresh or dried root tea; flower essence. I take a dropper and collect the water held within the leaves in a bottle, then later add a few drops to a flower essence or tincture blend.

Some additional garden practices

❦ Have a vision for your garden and yourself as the gardener. Dream big but start smaller than that. Keep the soil's health in the forefront of your mind.
❦ Take time to sit quietly and walk in and around the edges of your garden. This regular practice will help you establish a connection with nature's rhythms and strengthen your inner and outer abilities to be more fully present in your daily life.

Cutting and washing teasel roots

BOTTOM LEFT
A few purple teasel flowers encircle the plant's cone head.

🌿 Pay attention to the movement of the sun and moon through each month and season, noticing their light and shadows, their ebbing and flowing rhythms, their ever-changing qualities. Contemplate the dynamic forces influencing nature from the larger cosmos.

🌿 Avoid compacting the soil of your garden beds. Remind your human friends to walk on the paths and avoid stepping directly onto a bed. Keep all soil covered with plants or mulch throughout the year. Dance, sing, and pray in your garden. Learn to laugh and leap over beds whenever the opportunity arises. Create special altars for contemplation and prayer and for giving offerings to the Ancestors.

I understand *healing* as a verb, a lifelong practice of conscious participation. Other verbs that relate to healing include *restoring*, *rebuilding*, *regenerating*, *reconciling*, *mending*, and *repairing*. Herbs are friends to call upon for support, as they are deeply connected to the wisdom of the Earth and the cosmos, embodying a spiritual consciousness and intelligence that guides the healing process.

Whether you call yourself an herbalist, a gardener, or a person interested in herbs, may this book offer inspiration and information that benefits you, your community, and our Earth. Cultivating herbs without chemicals, and gathering and preparing them with gratitude and love, are life-affirming acts. May herbal remedies find their rightful place in every household and community clinic, not as commodities but as beloved friends.

CLOCKWISE FROM TOP LEFT
A monarch on goldenrod

Passing on my knowledge is a large part of my work; I lead monthly Herb Walks through Avena's gardens each summer to engage with the local and visiting communities about herbalism and plants.

Smelling fresh lavender

Walking bridge through the summer garden

Acknowledgments

What gets me out of bed early each morning are the healing plants and pollinators and my commitment to serve them. My gratitude for these plants, their pollinators, and the people who have kept herbal medicine alive throughout the centuries is immense and beyond words.

A huge bow to Molly Haley, whose exquisite photographs give life, meaning, and soul to this book. Collaborating with you has been a joyful honor. May your beautiful images spread far and wide, uplifting people's hearts and minds.

So much gratitude to the following women whose teachings and lives show me how to embody presence and walk upright. My respect for each of you is enormous: Rocio Alarcón Gallegos, Claudia Ford, Karyn Sanders, Sarah Holmes, Jennifer Neptune, Sherri Mitchell, Miigam'aghan, Jus Crea Giammarino, Rachel Bagby, Rowen White, Amanda David, Mandana Boushee, Leah Penniman, Stephanie Morningstar, Laura Riccardi-Lyvers, Thea Marie Carlson, Kate Gilday, Kay Parent, Amy Goodman, Laura Lamb Brown-Lavoie, Luisa Acosta, Mercedes Santiago, Rosemary Gladstar, Charis Lindrooth, Squidge Davis, Meredith Bruskin, Priscilla Skerry, Amy Jenner, Tonya Lemos, Molly Gawler, Elsie Gawler, and the late Marija Gimbutas.

Huge appreciation to past, present, and future Avena Botanicals gardeners, staff, students, and Herbal Classroom board members for your hard work, dedication, and creativity. Farming is demanding work. Navigating government regulations is exhausting and frustrating. Thank you everyone, especially Julia Yelton, Bob Harden, Kiera MacKenzie, Stephanie Cesario, Zoe Hayes, Brittany Cooper, Elsie Gawler, Anna Leavitt, Katie Ring, Anna Lataille, Amanda Affleck, Johnny Affleck, Sara Vesta, Natalie Travia, Erda Grass, Alyson Fleming, Audrey Maddox, Robin Storm, Michelle Cross, Pettina Harden, Katie McCormick, Caitlin Boggs, Pat Conant, Kelly Goodson, and Viennia Booth for the numerous ways you have contributed to helping medicinal herbs and healing gardens be a possibility in today's world. Much gratitude to the various gardeners who contributed to organizing Avena's drying room procedures and logbooks, including Lauren Cormier, Denise DeSpirito, Stephanie Cesario, Zoe Hayes, Elise Gawler, and Brittany Cooper. And a big bow to herbalist Suzanna Stone, for inviting me to teach at an herbal symposium titled Reciprocity of Pollination—these words inspire me daily.

Special thanks to three long-term friends, Russel Comstock, Katey Branch, and the late Alan Day, who believed in my vision for Avena Botanicals and helped me and Avena's farm grow into who we are today, and to three other long-term friends, Selkie O'Mira, Abe Baggins, and Melissa Hatch, whose support, humor, and love mean so much. Big thanks to Spencer, organic dairy farmer, who for many years has allowed two large dump trucks of cow manure to leave his farm and come to ours. Avena's compost is magnificent because of your cows!

Thank you, Jan Hartman, for asking me to write this book for Princeton Architectural Press. Your caring words and support live in my heart. To Sara Stemen, editor at Princeton Architectural Press, and copyeditor Lindsey Westbrook, a big thank-you for your skills and support. Thank you Natalie Snodgrass for the beautiful and graceful way you designed this book. Thank you also to all the staff working behind the scenes, including Lynn Grady, Janet Behning, Valerie Kamen, and Paul Wagner. I hope someday you will come visit my garden so I can thank you in person.

To my mother and father, Pat Soule and Lester Soule, my grandmother Katherine Soule, and my sister, Lisa Wind, thank you for your love and support. For my partner, Tom Griffin, who quietly feeds me and dozens of families fresh, biodynamic food, huge hugs of gratitude. To my ancestors, known and unknown, I humbly acknowledge your presence. Thank you, Karl, and the Ancestors of this land who allow us to be here and who generously offer protection, guidance, and love.

For over twenty-five years, I have intentionally integrated flowers for hummingbirds into Avena's garden. These tiny birds, weavers of light, teach humans about pollination, ecology, beauty, healing, perseverance, and so much more. The ruby-throated hummingbird, the only hummingbird species that migrates to Maine in summer, travels through several different Black, Brown, and Indigenous peoples' communities in their three-thousand-mile migration. May the people, plants, and pollinators along these migration corridors be protected and respected, always.

The following words from Terry Tempest Williams's book *Erosion* leave me feeling humbled and unsettled: "How do we hold ourselves to account over our inescapable complicity in a fossil fuel economy that is contributing to climate change, as well as ravaging tribal and 'public' lands? What are the necessary actions we can take in order to realize justice for all? And how do we find the strength to not look away from all that is breaking our hearts?" Thank you, Terry, for your essays of undoing.

And thank you to people everywhere who are working tirelessly to end the violence that for centuries has affected Earth and all living beings. May the medicine plants find their way to you, comforting your hearts and healing your spirits.

Resources

Social, Racial, and Environmental Justice

—

Books

Bowen, Natasha. *The Color of Food: Stories of Race, Resilience and Farming*. Gabriola Island, BC: New Society, 2015.

brown, adrienne maree. *Emergent Strategy: Shaping Change, Changing Worlds*. Chico, CA: AK, 2017.

DiAngelo, Robin. *White Fragility*. Boston: Beacon, 2018.

Dunbar-Ortiz, Roxanne. *An Indigenous Peoples' History of the United States*. Boston: Beacon, 2014.

Heller, Laurence, and Aline LaPierre, *Healing Developmental Trauma: Regulation, Self-Image, and the Capacity for Relationship*. Berkeley, CA: North Atlantic, 2012.

Kendi, Ibram X. *How to Be an Antiracist*. New York: One World, 2019.

King, Ruth. *Mindful of Race: Transforming Racism from the Inside Out*. Boulder, CO: Sounds True, 2018.

Lorde, Audre. *Sister Outsider*. Berkeley, CA: Crossing, 1984.

Menakem, Resmaa. *My Grandmother's Hands: Racialized Trauma and the Pathway to Mending Our Hearts and Bodies*. Las Vegas, NV: Central Recovery, 2017.

Mitchell, Sherri. *Sacred Instructions: Indigenous Wisdom for Living Spirit-Based Change*. Berkeley, CA: North Atlantic, 2018.

Oluo, Ijeoma. *So You Want to Talk about Race*. New York: Seal, 2018.

Ortiz, Paul. *An African American and Latinx History of the United States*. Boston: Beacon, 2018.

Penniman, Leah. *Farming While Black*. White River Junction, VT: Chelsea Green, 2018. See especially the chapter "White People Uprooting Racism."

Penniman, Naima, and Alixia Garcia. *Climbing Poetree*. Seattle: Whit, 2014.

Shiva, Vandana. *Earth Democracy: Justice, Sustainability, and Peace*. Berkeley, CA: North Atlantic, 2015.

———. *Soil Not Oil: Environmental Justice in an Age of Climate Crisis*. Berkeley, CA: North Atlantic, 2015.

Treleaven, David. *Trauma-Sensitive Mindfulness: Practices for Safe and Transformative Healing*. New York: Norton, 2018.

—

Websites and podcasts

1619 (podcast)
www.nytimes.com/2020/01/23/podcasts/1619-podcast.html
New York Times series on how slavery has transformed America

Agricultural Justice Project
www.agriculturaljusticeproject.org
Certification and technical assistance for farms and food businesses

Beyond Diversity 101
www.bd101.org
Workshops and community forum

For the Wild (podcast)
forthewild.world/podcast
An anthology of the Anthropocene hosted by Ayana Young

Grow a Row of Calendula
www.helpgrowarow.org
See page 119.

Healing Turtle Island
www.healingturtleisland.org
Repairing the history of violence

Maine-Wabanaki REACH
www.mainewabanakireach.org
Support for the Wabanaki community

Northeast Farmers of Color Land Trust
nefoclandtrust.org
Support for the work of QTBIPOC farmers, land stewards, and Earth workers

On Being with Krista Tippett (podcast)
Episode: "Ruby Sales: Where Does It Hurt?" (September 15, 2016),
www.onbeing.org/programs/ruby-sales-where-does-it-hurt
An interview with civil rights legend Ruby Sales

People's Institute for Survival and Beyond
www.pisab.org
Community-organizing workshops for social transformation

Resource Generation
www.resourcegeneration.org
A multiracial membership
community committed to the
redistribution of wealth, land,
and power

Rootwork Herbals
www.rootworkherbals.com
Herbalists Amanda David and
Mandana Boushee offer Woke
without the Work, an online class
for non-BIPOC herbalists.

Sacred Instructions
www.sacredinstructions.life
Website of Indigenous activist
Sherri Mitchell

Soul Fire Farm
www.soulfirefarm.org
A BIPOC-centered community
farm committed to ending racism
and injustice in the food system

Herbal and Biodynamic Education, Herbal Remedies, Gardening, Ecology, and Meditation

—

Books and printed matter

Alarcón Gallegos, Rocio. *Messages from the Hummingbirds*. Hand-painted oracle cards with Spanish and English descriptions, available in the United States through www.avenabotanicals.com. For information on Rocio's work, see www.iamoe.org.

Carlson, Thea Maria. "Awakening the Heart in Agriculture." Lecture presented at the International Biodynamic Conference in Dornach, Switzerland, February 8, 2020. www.sektion-landwirtschaft .org/en/lwt/2020/einzelansicht /awakening-the-heart-in -agriculture-by-thea-maria -carlson.

Carpenter, Jeff, and Melanie Carpenter. *The Organic Medicinal Herb Farmer: The Ultimate Guide to Producing High-Quality Herbs on a Market Scale*. White River Junction, VT: Chelsea Green, 2015.

Dalton, David. *Stars of the Meadow: Medicinal Herbs as Flower Essences*. Herndon, VT: Lindisfarne, 2006.

Estabrook, Lisa. *Tending the Garden of Your Soul: Soulflower Plant Spirit Wisdom and Oracle Guidebook*. Yarmouth, MA: Soulflower Plant Spirit Art, 2019. Available at www .mysoulflower.com, along with hand-painted Soulflower Plant Spirit Oracle Deck and journal.

Geniusz, Mary Siisip. *Plants Have So Much to Give Us, All We Have to Do Is Ask: Anishinaabe Botanical Teachings*. Minneapolis: University of Minnesota Press, 2015.

Geniusz, Wendy Makoons. *Our Knowledge Is Not Primitive: Decolonizing Botanical Anishinaabe Teachings*. Syracuse, NY: Syracuse University Press, 2009.

Gladstar, Rosemary. *Rosemary Gladstar's Medicinal Herbs: A Beginner's Guide*. North Adams, MA: Storey, 2012.

Hauk, Gunther. *Toward Saving the Honeybee*, 2nd ed. Great Barrington, MA: Steiner, 2017.

His Holiness the Dalai Lama and Archbishop Desmond Tutu. *The Book of Joy*. New York: Avery, 2016.

Hughes, Nathaniel, and Fiona Owen. *Weeds in the Heart*. New Maldon, UK: Quintessence, 2016.

Johnson, Wendy. *Gardening at the Dragon's Gate*. New York: Bantam, 2008.

Kimmerer, Robin Wall. *Braiding Sweetgrass: Indigenous Wisdom, Scientific Knowledge, and the Teachings of Plants*. Minneapolis: Milkweed Editions, 2013.

Kornfield, Jack. *A Path with Heart: A Guide through the Perils and Promises of Spiritual Life*. New York: Bantam, 1993.

Levy, Juliette de Bairacli. *Common Herbs for Natural Health*, rev. ed. Woodstock, NY: Ash Tree, 1996.

Manuel, Zenju Earthlyn. *The Way of Tenderness: Awakening through Race, Sexuality, and Gender*. Boston: Wisdom, 2015.

McBride, Bunny. *Women of the Dawn*. Lincoln, NE: Bison, 2001.

McIntyre, Anne. *Herbal Treatment of Children: Western and Ayurvedic Perspectives*. London: Elsevier, 2005.

Nabhan, Gary Paul. *Cross-Pollinations: The Marriage of Science and Poetry*. Minneapolis: Milkweed Editions, 2004.

National Research Council of the National Academies. *Status of Pollinators in North America*. Washington, DC: National Academy of Sciences, 2007.

Nhat Hanh, Thich. *The Miracle of Mindfulness: A Manual on Meditation*. Boston: Beacon, 1975.

Phillips, Michael. *Mycorrhizal Planet: How Symbiotic Fungi Work with Roots to Support Plant Health and Build Soil Fertility*. White River Junction, VT: Chelsea Green, 2017.

Soule, Deb. *How to Move Like a Gardener*. Rockport, ME: Under the Willow, 2013, 2016.

———. *Healing Herbs for Women*. New York: Skyhorse, 2016.

Thanissara. *Time to Stand Up: An Engaged Buddhist Manifesto for Our Earth*. Berkeley, CA: North Atlantic, 2015.

Thornton Smith, Richard. *Cosmos, Earth and Nutrition: A Biodynamic Approach to Agriculture*. East Sussex, UK: Sophia, 2009.

Turner, Nancy J. *Ancient Pathways, Ancestral Knowledge: Ethnobotany and Ecological Wisdom of Indigenous Peoples of Northwestern North*. Montreal, QC: McGill-Queen's University Press, 2014.

Weininger, Radhule. *Heartwork: The Path of Self-Compassion*. Boulder, CO: Shambhala, 2017.

Whitten, Greg. *Herbal Harvest: Commercial Organic Production of Quality Dried Herbs*. Toorak, VIC: Bloomings, 1999.

Wildfeuer, Sherri, *Stella Natura Biodynamic Planting Calendar*. Available at www.stellanatura.com.

—

Websites and podcasts

Ancestral Apothecary School
www.ancestralapothecaryschool.com
School of herbal, folk, and Indigenous medicine

Avena Botanicals Herbal Apothecary
www.avenabotanicals.com;
www.youtube.com:
AvenaBotanicalsvideo
Herb walks and talks

Biodynamic Farming & Gardening Association
www.biodynamics.com
Training, research, conferences, and online courses

Blue Otter School of Herbal Medicine
www.blueotterschool.com
Consultations and classes

Eden Land Farm
www.dp3herbs.org
Black-owned and led medicinal herb farm in Gainesville, AL, run by Dr. Marlo Paul and Dr. Anthony Paul

Harriet's Apothecary
www.harrietsapothecary.com

Healing Spirits
www.healingspiritsherbfarm.com
Organic herbs grown in New York State

The Herbal Classroom
www.avenabotanicals.com/pages/herbal-classroom
Classes hosted by Avena Botanicals Healing Gardens

The Herbal Highway (podcast)
kpfa.org/program/the-herbal-highway/
From California radio station KPFA, hosts Karyn Sanders and Sarah Holmes covering herbal medicine and promoting Indigenous land rights.

Josephine Porter Institute
www.jpibiodynamics.org
Biodynamic preparations and books

Maine Organic Farmers & Gardeners Association
www.mofga.org
Advocacy, support, and policy development for farmers

Pacific Botanicals Herb Farm
www.pacificbotanicals.com
Organic, medicinal herb farm and online shop

Queering Herbalism,
     Herbal Freedom School
     queerherbalism.blogspot.com
     School and community supporting
     BIPOC herbalists
Spikenard Farm Honeybee Sanctuary
     www.spikenardfarm.org
     Virginia sanctuary for sustainable
     and biodynamic beekeeping
Woodland Essence
     woodlandessence.com
     Source for essences, oils, salves,
     and other herbal products
Zack Woods Herb Farm
     www.zackwoodsherbs.com
     Organic herbs grown in Vermont

Seeds and Tools
—

Fedco Seeds
     www.fedcoseeds.com
     Source for hori-hori, EZ-Digger,
     Felco pruners and clippers,
     Reemay, gloves, collapsible drying
     racks
Green Heron Tools
     www.greenherontools.com
     Source for *kama*
High Mowing Organic Seeds
     www.highmowingseeds.com
Indigenous Seed Initiative
     www.indigenousseedinitiative.org
Johnny's Selected Seeds
     www.johnnyseeds.com

Kinsman Company
     www.kinsmangarden.com
     Source for heat-treated carbon
     steel tools with wooden handles:
     border forks, border spades,
     digging forks, digging spades
Maine Indian Basketmakers Alliance
     www.nativeartsandcultures.org
     /maine-indian-basketmakers
     -alliance-miba
Sierra Seeds
     www.sierraseeds.com
Southern Exposure Seed Exchange
     www.southernexposure.com
Strictly Medicinal Seeds
     strictlymedicinalseeds.com
Terrain
     www.shopterrain.com
     Source for Clarington Forge tools:
     border forks and spades
Truelove Seeds
     www.trueloveseeds.com
Turtle Tree Biodynamic Seed Initiative
     www.turtletreeseed.org
United Plant Savers
     www.unitedplantsavers.org

Index

Deb Soule (right) was raised in a small town in western Maine. Her sensitivity and curiosity to plants and birds began at a young age, when her grandmother often took her to special places to show her spring flowers and birds. At age sixteen, Soule began organically gardening and studying medicinal plants. In 1980, she spent four months living in Nepal, near three Tibetan monasteries, and was deeply moved by the Tibetan people's commitment to easing physical ailments and mental and emotional challenges with plants, prayer, and other spiritual practices. Five years later, in 1985, Soule founded Avena Botanicals Herbal Apothecary to share high-quality herbal medicine with women living in rural areas. In 1995, Avena Botanicals moved to its current farm and Soule began designing and building medicinal herb gardens, practicing biodynamic agriculture, and teaching classes. In 2017 and 2020, she spent time in Ecuador with the ethnobotanist and healer Rocio Alarcón Gallegos, studying hummingbirds, which have always fascinated her. Soule is the author of *How to Move Like a Gardener* and *Healing Herbs for Women*. She is the founder of the grassroots initiative Grow a Row (helpgrowarow.org), inspiring gardeners to grow a row of calendula, dry the flowers, create oils and salves, and donate them to projects that serve women and children who have experienced domestic violence and sex trafficking.

Molly Haley (left) is a freelance photographer based in Maine, specializing in portraiture and documentary photography. After graduating with a bachelor's degree in Spanish, she moved to Portland, Maine, to study photojournalism, radio, and videography at the Salt Institute for Documentary Studies. Haley then went on to work as the director of multimedia at the Telling Room, a nonprofit creative writing center. There, she taught immigrant and refugee teens tools to share their stories with their community. Haley has traveled internationally to photograph for nonprofits in Nepal, Honduras, Guatemala, and the Dominican Republic. As a photojournalist, Haley has been published in *O, the Oprah Magazine*; the *Sun Magazine*; the *Atlantic*; *Down East* magazine; *Maine Women* magazine; *Maine Farms*; and others. More of Haley's photographs can be seen at www.mollyhaley.com.

Published by
Princeton Architectural Press
202 Warren Street
Hudson, New York 12534
www.papress.com

Photography: Molly Haley
Editors: Jan Hartman and Sara Stemen
Designer: Natalie Snodgrass

Library of Congress Cataloging-in-Publication Data
Names: Soule, Deb, author.
Title: The healing garden : herbs for health and wellness / Deb Soule.
Description: Hudson, New York : Princeton Architectural Press, [2021] |
    Includes bibliographical references. | Summary: "A guide to growing and harvesting
    healing plants, with advice on how to create herbal preparations and in-depth
    discussions of the properties and uses of various plants"—Provided by publisher.
Identifiers: LCCN 2020028939 | ISBN 9781616899264 (hardcover)
Subjects: LCSH: Herb gardening. | Herbs—Therapeutic use. | Medicinal plants.
Classification: LCC SB351.H5 S678 2021 | DDC 635/.7—dc23
    LC record available at https://lccn.loc.gov/2020028939